Entropic Invariants of Two-Phase Flows

Entropic Invariants of Two-Phase Flows

By

Eugene Barksy
Azrieli College of Engineering, Jerusalem

ELSEVIER

AMSTERDAM • BOSTON • HEIDELBERG • LONDON
NEW YORK • OXFORD • PARIS • SAN DIEGO
SAN FRANCISCO • SINGAPORE • SYDNEY • TOKYO

Elsevier
Radarweg 29, PO Box 211, 1000 AE Amsterdam, The Netherlands
The Boulevard, Langford Lane, Kidlington, Oxford OX5 1GB, UK
225 Wyman Street, Waltham, MA 02451, USA

Notices

Knowledge and best practice in this field are constantly changing. As new research and experience broaden our understanding, changes in research methods, professional practices, or medical treatment may become necessary.

Practitioners and researchers must always rely on their own experience and knowledge in evaluating and using any information, methods, compounds, or experiments described herein. In using such information or methods they should be mindful of their own safety and the safety of others, including parties for whom they have a professional responsibility.

To the fullest extent of the law, neither the Publisher nor the authors, contributors, or editors, assume any liability for any injury and/or damage to persons or property as a matter of products liability, negligence or otherwise, or from any use or operation of any methods, products, instructions, or ideas contained in the material herein.

Library of Congress Cataloging-in-Publication Data

Barsky, Eugene, 1974- author.

Entropic invariants of two-phase flows / Eugene Barksy, Azrieli College of Engineering, Jerusalem, Israel. – First edition.

pages cm

ISBN: 978-0-12-801458-5

1. Two-phase flow. 2. Statistical mechanics. 3. Computational fluid dynamics. I. Title.

TA357.5.M84B374 2014

620.1'064–dc23

2014017896

British Library Cataloguing in Publication Data

A catalogue record for this book is available from the British Library

ISBN: 978-0-12-801458-5

This book has been manufactured using Print On Demand technology. Each copy is produced to order and is limited to black ink. The online version of this book will show color figures where appropriate.

Contents

Introduction vii

CHAPTER 1 Modern Conceptions of Thermodynamic Entropy 1

1. Entropy substantiation 1
2. Entropy, ensemble of states, probability 7
3. Problematic aspects of entropy 10

CHAPTER 2 Invariants for Continuous Flows 17

1. Substantiation of the necessity to model technological processes 17
2. Certain aspects of continuous flows modeling 18
3. Invariants for continuous flows 22
4. Analysis of flow parameters using similarity criteria 28

CHAPTER 3 Modern Notions of Two-Phase Flows 37

1. Peculiarities of unit particles settling 37
2. Final settling velocities 43
3. Particle interaction with a moving medium 46
4. Features of two-phase flow generation 49
5. Mass suspension of particles in a flow 52
6. Carrying capacity of two-phase flows 56

CHAPTER 4 Empirical Invariants for Two-Phase Flows 59

1. Fractional extraction 59
2. Affinization of fractional extraction curves 64
3. Unified separation curves 68
4. Generalizing empirical invariant 74

CHAPTER 5 Entropy of Mixture Composition 79

1. Invariants for granulometric composition 79
2. Uncertainty of mixture composition 83

CHAPTER 6 Two-Phase Flow as a Statistical System 91

1. Substantiation of physical analogy 91
2. Model of particle collisions in a two-phase flow 94
3. External interaction model in a two-phase flow 98
4. Statistical model substantiation 101

CHAPTER 7 Main Statistical Parameters of a Two-phase Flow 105

1. Substantiation of a statistical system for solid particles continuum 105
2. Generalising parameter for a two-phase flow 106
3. Generalising invariant for two-phase flows 113
4. Definition of chaotising factor for two-phase flows 117

CHAPTER 8 Substantiation of Statistical Parameters of Mass Transfer in Two-Phase Flows **121**
1. Multidimensional model of a two-phase system 121
2. Two-phase flow mobility in separation regimes 126
3. Generalising invariants for parameters characterising two-phase flows 130
4. Substantiation of entropy of two-phase flows with a polyfractional solid phase 133
5. Equilibrium state as a condition of maximal entropy 140

CHAPTER 9 Specific Properties of Entropy of Two-Phase Flows **143**
1. Composition entropy as a criterion of solid-phase separation quality 143
2. Statistical equilibrium for particle distribution in a flow 148
3. Stability of entropy and the entire mass distribution process in a two-phase flow 155

CHAPTER 10 Invariants for Separation Curves **161**
1. Regimes of solid-phase particle motion in flows 161
2. Particle extraction in two-phase flow separation regimes 167

CHAPTER 11 Basic Physics of Cascade Separation **175**
1. Main principles of cascade mass-exchange processes 175
2. Discrete model 180
3. Analysis of a mathematical model of a regular cascade 187
4. Separation at cyclic feed of initial material into cascade apparatus 191
5. Absorbing Markov chain in cascade classification of bulk materials 193
6. Mathematical model of cascade separation with arbitrary distribution coefficients 195
7. Discrete model of critical regimes of vertical two-phase flows 200
8. Optimisation of principal parameters of multi-stage separation 212

CHAPTER 12 Application of the Obtained Results **223**
1. Invariants for estimating the perfection of the design of separation apparatuses 223
2. Method of theoretical estimation of vertical cascade separators 230
3. Experimental check of the adequacy of the method of cascade separators computation 232
4. Computation method of multirow combined separation cascade 235
5. Optimization of multirow classifier operation 239
6. Separator efficiency rank 242
7. Flow velocity ensuring optimal separation 243

Annotation 247
Bibliography 249
Index 251

Introduction

Two-phase flows are widely used in industry. In the overwhelming majority of cases, various technological processes involving such flows are organized in such a way that the carrier fluid flow is directed bottom-up. Separation regimes of such flows with one part of solid phase moving with the flow and another part moving counter the flow are rather widespread. For example, processes of minerals concentration, bulk materials fractionating by particles sizes, processes in boiling bed, powder materials drying, most mass-exchange processes are organized in this way.

Visual observations and rapid filming of two-phase flows with a poly-fractional solid phase in a transparent vertical channel show an absolutely chaotic motion of solid particles. They move upwards and downwards, to the left and to the right, take part in vortical motions of the carrier medium, collide with each other and with channel walls, form aggregates that permanently appear and decompose. All this makes it impossible to predict the direction and velocity of even one particle, not to speak about the behaviour of solid phase on the whole.

This can explain why scientists have not attained theoretical understanding of such flows until now. It is not surprising, because until today there is no acceptable general theory even for purely one-phase liquid and gas flows. It is known that many prominent specialists have been working for many years in this field in many countries around the world. Numerous theoretical schemes for pure turbulent flows have not led to generalizing results, but only to a grim joke saying that 'turbulence is a cemetery of theories'.

However, at the same time, dams, hydroelectric power station, pneumo- and hydro-transport systems, and more are being built and operated. They all successfully function and practically always reach their design characteristics.

As known, it is a result of fragmentary, and not general regularities of flows, which have not been combined into a unified theory yet. Such regularities fix only some 'patch-like' particulars of flows. For one-phase flows, such fragmentary regularities include so-called similarity criteria derived from rather simple mathematical models. At the same time, these criteria possess an explicit physical meaning. In hydrodynamics, they are, as a rule, dimensionless, and each of them consists of parameters reflecting some characteristic features of flows. These parameters can acquire an infinitely large number of numerical values, while the criterion can remain constant.

Such relations are usually called invariants. It is a rather broad notion, which can be defined as follows:

Invariant is a system of values, parameters, or regularities describing a group of phenomena remaining unchanged at one or another transformation of parameters inside this system.

It is noteworthy that invariants are not obligatorily dimensionless. Even graphical relations can serve as invariants at a respective choice of coordinates, so that all experimental or estimated values of parameters fitted the same curve.

It is very important to study the possibility of finding such invariants for two-phase flows, which are, for good reason, much more complicated than one-phase flows. This will make it possible to start the development of a general approach to the construction of their theory.

The initial study of the process started with the study of the behaviour of solitary particles. First publications on the topic appeared in the second half of the nineteenth century. First, regularities of settling of solitary spherical particles in an unlimited motionless liquid were studied. In numerous subsequent papers devoted to the study of this phenomenon, the effect of various factors (fluid and material densities, final velocities of particles settling, their drag coefficients, etc.) on these regularities was revealed. A transition to materials with irregular particles shape has been so complicated that publications dealing with these problems are still appearing nowadays. An attempt of applying main regularities obtained when studying the behaviour of solitary particles in moving medium to real processes has not given positive results.

Today all the problems of two-phase flows are solved at an empirical level. Numerous researches are continuously performed on models of actual industrial units. The transfer of the obtained results onto pilot plants is not always successful. A huge volume of experimental studies has allowed the development of numerous empirical computation methods for specific apparatuses. However, the best and the most grounded empirical formula can be applied with satisfactory results only within a limited range determined by the conditions of its derivation. Frequently used extrapolation of such dependencies beyond the experimental range limits leads to bad mistakes. Besides, empirical relations do not reflect the regularities of the phenomena under study, but only provide their quantitative characteristics. Any carefully conducted experiment does not allow taking into account the diversity of constant and random factors—both quantitative and qualitative ones. The latter can be taken into account least of all.

Since it is unrealistic at present to discuss the possibility of developing a general theory of two-phase flows, the efforts should be concentrated on the search for physically grounded invariants for them. The chaotic disorder in the motion of solid particles in two-phase flows by no means points to the absence of general regularities. On the contrary, rigid internal regularities can manifest themselves in such systems only through a general chaotization, as it was established, for example, during the development of kinetic theory of gases. This theory was developed within the framework of thermodynamics. To understand the logics and the essence of the approach to the problem that led to the development of the kinetic theory of gases, we will try to understand how the development of thermodynamics led to this theory.

Initially it was known that fire releases heat that can lead to an increase in the volume of any bodies—solid, liquid, and gases. As a result, work is performed. Fire led people to a steam engine, which became the basis of the industrial society development in the beginning of the nineteenth century. To create steam engines, it was necessary to establish the connection between heat and work. In search of a solution to this problem, energy conservation law (the first law of thermodynamics) was formulated. It implies that heat is of the same nature as energy. In a heat-engine, heat is converted into work, but the amount of energy is conserved. The law of total potential and kinetic energy conservation was established much earlier, in classical mechanics developed by Newton, Lagrange, and Hamilton. It was outstandingly successful and is still developing at present. One of its distinctive features is a strict determinism.

Classical mechanics can predict the behaviour of a system with either a small number of elements or with many symmetrically arranged elements. If the initial conditions for a system are specified, and forces acting on it are determined, one can follow the changes in this system both in the future and in the past using classical mechanics.

An especially important conclusion of this mechanics was that due to Newton's works it has become clear that nature obeys simple universal laws that are cognizable and can be expressed using exact mathematical language. Since then, experiments, quantitative study of various physical values in their interconnection, mathematical relations between them have formed the basis for the insight into the secrets of nature.

Using classical mechanics, it has become possible to carry the problem of celestial bodies motion to its conclusion, but it proved to be absolutely helpless facing the problem of three bodies motion in a general case, not to speak about mass motion of many bodies.

In the seventeenth and eighteenth centuries, Newton's works had such a high prestige that insistent attempts were made to reduce all laws of nature, discovered by that time, to classical mechanics. In the long run, they were unsuccessful. However, at least in one field of science these attempts led to excellent results, namely, in the theory of heat phenomena, but it happened only in the middle of the nineteenth century. By that time, confidence in the molecular structure of substances started strengthening. Therefore, efforts were made to find out how macroscopic properties of a substance depend on the behaviour of imaginary molecules it is composed of. The first successes in this direction concerned gases in which molecules interact with a negligible force. The obtained results allowed expressing pressure, temperature, and other macroscopic parameters of gas through such an averaged characteristic of molecules as their kinetic energy. Therefore, this theory was called a kinetic theory or statistical mechanics. It is based on fundamental Boltzmann's works, where speculative analysis allowed establishing statistical relations between the molecular structure of substance and its microscopic properties at a microscopic level.

Statistical mechanics was based on classical mechanics, but it established new relations and introduced new concepts. A starting point for Boltzmann was atomistic theory stating that matter consists of immense number of small moving balls.

In principle, it is possible to describe the motion of a system of independent particles from the standpoint of classical mechanics. Writing a differential equation for each particle, one can obtain comprehensive information on the system motion. However, for a system with an immense number of particles N, it is necessary to solve in a general form $2N$ differential equations, which is practically unrealisable due to the impossibility to specify initial conditions for each particle. At first sight, we could conclude that with increasing number of particles, properties of the system under study should become incredibly more complicated, which excludes the possibility of finding some regularity therein. However, it is not true, since at a large number of particles, new peculiar relations arise, which by no means can be reduced to purely mechanical phenomena. Here qualitatively new regularities arise having characteristic of statistical mechanics only. Their main distinctive feature is the use of probabilistic notions based on the understanding of ambiguity, randomness, and uncertainty of mass events. Before the works of Boltzmann, physicists believed that probability and uncertainty lead to a total randomness and incognizability, i.e. to chaos. In fact, it has turned out that a strict scientific study of uncertainty of events is possible. It had started long before Boltzmann's works with the analysis of roulette wheel results carried out by Bernoulli and Pascal. This analysis required calculation of probabilities, which had not been considered serious mathematics for a long time.

The main idea of probability determination is usually illustrated by the fact that it is totally undeterminable whether we obtain heads or tails when tossing a coin. However, if we toss it very many times, overall occurrence of heads or tails approaches 50%. A similar result is observed at simultaneous tossing of a large number of identical coins. This transition from a complete indefiniteness to an almost exact definiteness in case of a long sequence of events or large systems is the main idea in the study of randomness.

Principal laws of classical mechanics describe reversible processes and do not imply the possibility of irreversibility. Time (t) appears in these laws in the second degree. Therefore, substitution of $-t$ for

$+t$ gives irreversibility. This allows us to estimate future in a respective problem with the same degree of certainty as the past.

The irreversibility revealed in thermodynamics was substantiated by statistical mechanics. The first step in irreversibility formalization was made when determining entropy. There exists a rule called the second law of thermodynamics. It reads that in any physical process entropy remains constant or increases, and if it increases, the process is irreversible. All this seemed rather mysterious and not very satisfactory at that time. What is the sense of entropy? Why does it always increase and not decrease?

Boltzmann tried to solve these problems on the basis of 'atomistic' hypothesis and obtained amazing results. They were as important for physics of the twentieth century as the discovery of relativity theory and quantum mechanics. During more than hundred recent years, his concepts have demonstrated their scientific power and are now applied to various situations far beyond the limits of physical problems they were meant for. To the first approximation, it is very popular to explain the notion of entropy by the following reasoning. We imagine gas atoms contained under normal conditions in a closed vessel with the volume of, say, one cubic meter. Atoms in this vessel are in a certain configuration with respect to each other and vessel walls, but they permanently move, and the system configuration permanently changes. It means that a system consisting of a multitude of particles can have a great number of configurations. For an external observer all these configurations look the same—as a cubic meter of gas. For an internal observer who is able to discern separate atoms, this volume of gas is rather ambiguous. Boltzmann defined the measure of this ambiguity as entropy parameter.

Our interest in everything stated above is due to a certain physically substantiated analogy observed between the kinetic theory of gases and two-phase flows. These flows can be considered as a mass process with a simultaneous participation of a large number of particles of various sizes. Although these two phenomena are essentially different, the analogy between them (at least in particulars) allows clarifying the general pattern of a two-phase flow. A broad range of various random factors (the flow turbulence, non-uniform pressure in the flow and solid phase concentration, shape, mass and dynamic drag of particles, their interaction with each other and with apparatus walls) affects this flow. All this predetermines markedly irreversible character of the two-phase flow. To clarify at least its most general regularities, it is necessary to use main ideas of statistical mechanics applying them, naturally, to specific conditions of the problem under study.

A fundamental distinction of this approach is that it is based on the necessity to determine the state of the entire system at once, irrespective of its size. Here the methods of mass processes analysis are understood as naturally statistical. Practice of mathematical modelling of technological processes shows that to construct a successful model, one does not have to aim at their thorough description. It is enough to find a simple scheme, if it reflects, to a sufficient degree, the essence of the modelled phenomenon.

At the present level of the knowledge of two-phase flows, substantiation of invariants connecting principal parameters characteristic of such flows may be considered a great success.

It is noteworthy that by now some invariants of separating flow regimes have been empirically established. Their checking in numerous experiments has shown that they are valid for two-phase flows with developed turbulence. Determining the mechanism of invariants formation for such flows will make it possible to deepen our insight and, possibly, provide a key to finding invariants for laminar and transient regimes. It is very important to understand how invariants of one-phase flows are formed.

This problem and modelling methods are examined in detail in this book. Since the statistical definition of entropy was formulated in 1872, it has been considered not only as a sign of irreversibility, but also as a measure of disorder or uncertainty of the respective systems.

One can imagine that a complicated dynamic system with a large number of objects, whose characteristics allow their interpretation through probabilities, should generate entropy. This idea is widely used in modern science in many fields far from thermodynamics, whose main objects are mass systems. As a rule, entropy plays the role of the main numerical invariant of transformations in the theory of these systems. Therefore, it is necessary to know the ways of this parameter formation in thermodynamics, to understand its properties and clarify some details.

CHAPTER

1 Modern Conceptions of Thermodynamic Entropy

CHAPTER OUTLINE

1. Entropy substantiation 1
2. Entropy, ensemble of states, probability 7
3. Problematic aspects of entropy 10

1. Entropy substantiation

A function introduced by Clausius, in 1864, which he named entropy, obtained unexpected resonance in science. It appeared in the course of his attempts to understand the laws of heat transfer and conversion into other kinds of energy and had a rather simple form:

$$H = \frac{\Delta Q}{T} \tag{1.1}$$

where ΔQ is heat consumption and T is the absolute temperature of the consumption process.

Clausius introduced this function to characterise imperfections of the Carnot cycle. While thinking about the principles forming the basis of steam engine operation, Carnot concluded about a basic limit of work performed by this engine at the expense of a certain amount of heat. He found a dependence on temperatures causing the heat flux and independence of the engine design or universality and on the way of performing work. The main idea developed by Carnot is the notion of a cycle; in the course of its operation, heat engine passes through a cycle of states. An engine performing mechanical work owing to a heat flux returns to its initial state each time.

It was necessary to find a way of expressing the difference between useful energy exchanges in this cycle and irreversibly lost dissipative energy. Examining entropy increment dH during a short time interval dt, he separated it into two parts:

$$dH = d_eH + d_iH.$$

The value d_eH is due to an external cause – mass exchange with the environment. This part of entropy is reversible, which means that it can be both positive and negative.

In real heat engines, irreversible processes of heat loss and friction take place, side by side with reversible heat exchange. They lead to entropy increase, and this part of entropy d_iH is internal with respect to the system and never changes sign. In the long run, it reflects irreversible changes inside the system, which corresponds to a spontaneous evolution of the system. This refers to such processes

Entropic Invariants of Two-Phase Flows. http://dx.doi.org/10.1016/B978-0-12-801458-5.00001-0

inside the system as heat conductivity or diffusion. Under the influence of these processes, the system passes into an equilibrium state corresponding to the maximal entropy. It is accepted that entropy is the energy loss required for a reversible return of a system into its initial state after an adiabatic process completed at the temperature corresponding to said initial state. This point of view is retained until now and is reflected in most courses of thermodynamics. Another point of view states that entropy does not have the dimension of energy and reflects other features of a complicated system.

The main feature of this summarised function is that it changes in one direction only and always grows:

$$dH = d_e H + d_i H > 0.$$

In contrast to other physical quantities, such as temperature or pressure, entropy cannot be directly perceived or readily measured. It is determined by a mathematical expression and looks abstract and even artificial. Clausius attributed universal features to this parameter. He asserted that the entropy of the universe tends to the maximum, which gave rise to the "heat death" theory.

At first, this parameter was met with distrust, and, until now, heated debates about the physics of the notion of entropy arise repeatedly. Such a subtle and peculiar notion attracted many scientists and intellectuals, whose interests were sometimes far from thermodynamics, who tended to attribute a physical and even philosophical (not always righteous) meaning to entropy.

Today, the notion of entropy is referred not only to thermodynamic, but also to physicochemical or biological processes, crystallography, cosmology, black holes theory, as well as to information theory.

Clausius formulated the notion of entropy using macroscopic parameters only. His work completed the development of the theory of thermodynamics allowing the estimation of any processes connected to heat exchange. In addition, he contributed to the change in general ideas of the nature of heat by demonstrating that heat does not differ, in principle, from mechanical energy. He also suggested the idea that heat is realised in the form of kinetic energy of particles constituting a body under study.

After a time, this idea was developed by Boltzmann and allowed the revelation of the mechanism of the thermal laws of operation and peculiarities of entropy formation. Before discussing Boltzmann's ideas, it seems expedient to consider, in brief, the theory of heat transfer in gaseous systems. From the standpoint of the analysis of the revealed invariants, physics of these processes is very close to our problems.

One of the first quantitative laws describing the behaviour of gases was Boyle's law. It reads that at a fixed temperature T, gas volume V is inversely proportional to the pressure, as follows:

$$V = \frac{f(T)}{p}.$$

Much later, Charles formulated the law of gas volume dependence on temperature at a constant pressure, as follows:

$$\frac{V}{T} = \varphi(p).$$

Avogadro suggested a hypothesis that molar volumes of all gases contain equal numbers of molecules, as follows:

$$N_A = 6.022 \cdot 10^{23} \frac{1}{\text{mol}}.$$

Hence, gas volume at $T=\text{const}$ and $p=\text{const}$ is proportional to the number of moles n.

This relation for a gas can be written as follows:

$$\frac{pV}{T}=R \tag{1.2}$$

where R is an absolute gas constant. Thus, the absolute gas constant acquires the sense of an invariant for the three main parameters characterising the macrostate of a gaseous system. Not all the three parameters in Eq. (1.2) can be independent. Only two of them can be specified arbitrarily, while the magnitude of the third is connected to the absolute gas constant.

Gay-Lussac revealed that at the gas outflow into emptiness, the temperature remains unchanged because, in this case, work is not performed, and gas energy remains unchanged. It follows from the fact that the temperature remains unchanged, while the volume and pressure vary, that the energy of an ideal gas depends on temperature and is independent of the volume and pressure.

Hence, the energy of an ideal gas depends on its temperature and the number of moles, as follows:

$$U=nU_m(T)$$

where U_m is self-energy per mole and n is the number of moles.

The energy of a mixture of k components of an ideal gas equals a sum of the components' energies, as follows:

$$U=\sum_k n_k U_m(T).$$

Speaking about an ideal gas, we imply a gas of point particles having weight, but ideally having no size. Note that even in Boltzmann's works, non-zero particle size was accepted for the ideal gas, too.

For a mixture of ideal gases, Dalton's law is valid, which reads that the pressure generated by each component of the gas mixture is independent of other components and is described by the same law:

$$p_k U = n_k RT \tag{1.3}$$

where p_k is the partial pressure of the k-th component, and n_k is the number of moles of this component.

It is another invariant, whose analogy with Dalton's law is examined later.

Gay-Lussac experimentally proved that the thermal expansion coefficient of all gases at a constant pressure is approximately the same and equals $1/_{273}$ per centigrade.

Van der Waals suggested an ideal gas law taking into account molecule sizes and interactions between them.

According to the modern theory, mean kinetic energy per one particle is proportional to the temperature:

$$\frac{1}{2}m\overline{v^2}=\frac{3}{2}kT \tag{1.4}$$

where v is the mean molecule velocity, m is the molecule mass, and k is the Boltzmann constant. The gas law can be written in a simple form:

$$p=nkT. \tag{1.5}$$

Because Eq. (1.4) is independent of the kind of particles, the pressure per one particle is $\frac{kT}{V}$. To include the volume explicitly, we write Eq. (1.5) as:

$$pV = NkT. \tag{1.6}$$

Equations (1.5) and (1.6) are comparable with ideal gas laws of:

$$\begin{aligned}
&\text{Boyle} \quad pV = \text{const} \cdot (T = \text{const}) \\
&\text{Charles} \quad p = T \cdot \text{const} \\
&\text{Dalton} \quad p = nkT = \left(\sum_k n_k\right) \cdot kT = \sum p_k \\
&\text{Avogadro} \quad n = \frac{p}{kT}
\end{aligned} \tag{1.7}$$

(i.e., the number of particles in a unit gas volume depends on the pressure and temperature and is independent of the character of particles).

It is generally accepted that 1 g-mole of gas at the ice melting temperature (+273.15 K) occupies the volume 22.413 l. Thus, the Avogadro number is:

$$N_A = \frac{pV_{\text{gmol}}}{kT} = 6.0222 \cdot 10^{23}$$

whereas the Boltzmann constant is:

$$k = 1.38062 \cdot 10^{23}\,\text{J/K}.$$

The gas constant is determined as:

$$R = N_A \cdot k = 8.314\,\text{J/gmolK}.$$

Here, we recall one more regularity. In the early nineteenth century, Dulong and Petit derived an empirical rule reading that the specific molar heat capacity of atomic gases at the temperature above 100–200 K is independent of the system temperature and is constant for such gases. By way of example, atomic gases, for which this relationship is valid, are hydrogen (H) with the atomic weight taken as unity, nitrogen (N) with the atomic weight of 14 units, argon (Ar) – 40 units, iodine (I) – 126 units, and actinium (Ac) – 227 units, etc. Their heat capacity at a constant pressure is approximately equal to:

$$C_p \approx 21\,\text{J/Kmol}.$$

This is valid for all mentioned gases, although their molar masses are different. It is noteworthy that the specific heat capacity is based on an invariant allowing a certain expansion and clarification of the notion of entropy as determined by Boltzmann.

Nevertheless, the mechanism of energy absorption by molecules according to the theory of heat capacity advanced by Einstein as of yet remains unclear. It should be also noted that the entropy parameter introduced by Clausius is an invariant characterising a thermotechnical system in a specific way.

To understand the notion of the specific heat capacity of gas, it is necessary to examine the ratio between the quantity of heat and work. As known, any expanding system displaces surrounding bodies (i.e., executes a certain work). When an expanding gas displaces a piston by dh, it executes the work:

$$dA = F \cdot dh \tag{1.8}$$

where F is the force acting on the piston on the part of the gas:

$$F = p \cdot S \tag{1.9}$$

where p is the gas pressure and S is the piston area. Substituting Eq. (1.9) into Eq. (1.8), we obtain:

$$dA = pSdh = pdV \tag{1.10}$$

where V is gas expansion volume.

This is a simple, but important relationship.

During gas expansion from the volume V_1 to V_2, the executed work at $p = \text{const}$ is:

$$A = p(V_1 - V_2).$$

For $T = \text{const}$, for 1 g molecule:

$$p = \frac{RT}{V}.$$

Therefore,

$$dA = pdV = \frac{RT}{V}dV = RT \ln V.$$

At that, the executed work is:

$$A = RT \ln \frac{V_2}{V_1}.$$

If a system does not receive energy (heat) from the outside, the work is executed at the expense of the intrinsic energy of the system E:

$$dE = dQ - p \cdot dV \tag{1.11}$$

where dQ is the external energy (heat) and dE is the intrinsic energy, which is a function of state.

The relation of Eq. (1.11) expresses energy conservation law. If the body temperature grows by dT at the heat absorption dQ, then the ratio:

$$\frac{dQ}{dT} = C$$

is called the heat capacity of a body.

There exist heat capacity at a constant volume C_V and heat capacity at a constant pressure C_p. If in Eq. (1.11) $dV = 0$, then $dQ = dE$ (i.e., all the heat is consumed for increasing the intrinsic energy of a body), and we can write:

$$C_V = \left(\frac{dE}{dT}\right)_V. \tag{1.12}$$

If, however, $p = \text{const}$, the heat is consumed not only for increasing the intrinsic energy, but also for executing work:

$$dQ = dE + pdV = d(E + pV).$$

Hence,

$$Q = W = E + pV. \tag{1.13}$$

This function is called heat content or enthalpy. It is also a function of the body state. Hence, at $p = \text{const}$, the heat capacity is:

$$C_p = \frac{dW}{dT}.$$

It is clear from the comparison of Eqs (1.12) and (1.13) that:

$$C_p > C_V$$

always holds. It may seem that this inequality is connected only with the work that an expanding system should execute at its heating; however, it is not so. This inequality equally refers to certain bodies whose volume decreases with heating. Here, the so-called Le Chatelier principle becomes valid. Its main idea is that external impacts perturbing the thermodynamic equilibrium of a system cause processes tending to weaken the results of this impact.

Now we examine 1 g-molecule of a gas, whose respective heat capacities C_V and C_p are called molar ones. By virtue of the equation $pV = RT$, thermal function of 1 mol of gas is connected with its internal energy by the relationship:

$$W = E + pV = E + RT.$$

Differentiating this equality with respect to temperature, we obtain:

$$C_p = C_V + R$$

$$C_p - C_V = R$$

$$R = 8.3\ \text{J/mol} = 2\ \text{cal/mol}.$$

Monoatomic gas heat capacity can be readily found. In this case, the internal energy of gas is just a sum of kinetic energies of the translation movements of its particles. For one particle, this energy equals $\frac{3}{2}kT$. The internal energy of 1 mol of gas is:

$$E = \frac{3}{2}N_0kT = \frac{3}{2}RT.$$

Hence,

$$C_V = \frac{3}{2}R = 12.5\ \text{J/mol}$$

$$C_p = 20.8\ \text{J/mol}.$$

These magnitudes are temperature independent. We have recalled all this primary information for a better understanding of what follows.

2. Entropy, ensemble of states, probability

Classic thermodynamics deal with directly measurable quantities – pressure, temperature, volume, quantity of heat, etc., and represents a complete system. It allowed a complete and comprehensive control of various technological processes and industries. In the second half of the nineteenth century, it allowed a detailed estimation of furnaces, boilers, turbines, various heat exchangers, and, a bit later – steam, diesel, and petrol engines. These methods remain relevant until today and widely used without any special improvements. Actually, it was not a public necessity to advance the theory of these regularities at that time. However, scientists are always more interested in finding an answer to the question "why?" than "how much?" Despite the availability of perfect thermodynamics, this interest stimulated them to look into the details of its mechanism and, thus, to understand the causes of the observed phenomena. In the second half of the nineteenth century, attempts were made to reduce all natural phenomena to the laws of classic mechanics. Most of these attempts failed, but in the field of thermodynamics, thanks to the works of Boltzmann, this approach gave excellent results. This approach was based on heat identification with the motion of molecules. The simplest of these theories, which was developed first, was the kinetic theory of gases. It was a purely speculative model that could be devoid of physical significance, unless it was confirmed by the entire experimental material accumulated up to the present moment.

Today, it is hard to imagine the courage of the idea of identifying heat with the motion of atoms and molecules in the second half of the nineteenth century, when the existence of said atoms and molecules was not proved yet. This theory connects thermal energy with mechanical energy of material particles. This theory can be most visually applied to an ideal gas. Gas molecules are considered as vanishingly small balls moving at high linear velocities and permanently colliding with each other and with vessel walls. The kinetic energy of molecules macroscopically manifests itself as thermal energy of the gas under study. Thus, the entire thermal energy of gas is reduced to the kinetic energy of its molecules. However, it is impossible to take into account the influence of separate molecules on gas parameters. This influence can be an object of a kind of averaging (i.e., statistical generalisation).

The credit of successful solution of this problem should be given to Boltzmann. He introduced the notion of discontinuity and probability, which made a revolution in physics of that time. He established, from the standpoint of statistical approach, rather simple relations between the mechanical energy of molecules and all thermodynamic functions, in particular, entropy. This approach was developed by such representatives of new physics of the early twentieth century as Gibbs, Poincare, Lorentz, Plank, and Einstein.

The total result that can be considered as finally established is the fact that entropy of a certain state is connected with its probability. The connection between entropy and probability is accepted *a priori*, because these two quantities characterising a system always vary in the same direction. In fact, according to the Clausius principle, any system evolves in such a way that its entropy grows, and at the same time, such evolution is always directed to more probable states. This is usually illustrated by a simple example. We conceive a closed space (Figure 1.1) separated into two parts by an imaginary partition, and introduce a great number of gas molecules N, whose mean kinetic energy characterises

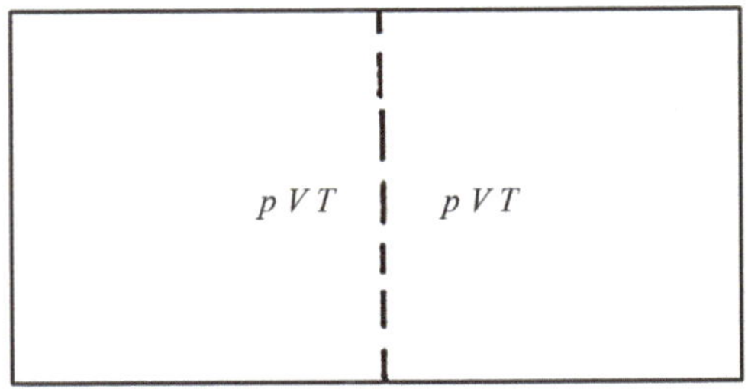

FIGURE 1.1 Chamber with a partition.

the gas temperature, into the left-hand part of the space. We let these molecules move spontaneously and look at their position after a certain period of time, when there are N_1 molecules in one part and N_2 in another. Obviously, each individual molecule has equal chances to be in both parts of the volume. In this case, the number of outcomes of the random value of the distribution, or the number of complexes, according to Landau, for $N = N_I + N_2$ is:

$$\varphi = \frac{N!}{N_1!N_2!}. \tag{1.14}$$

An ensemble or a complex of systems simultaneously represents all possible states of a system, whose commonness is because all of them consist of the same number of molecules N and possess the same definite energy.

Boltzmann, who was the first to discover the meaning of entropy as a measure of molecular chaos, concluded that the law of entropy growth reflects the growing disorganisation. According to such disorganisation, the most probable distribution corresponds to the approximate equality of the number of molecules in both parts (i.e., finally, to $N_1 = N_2 = \frac{N}{2}$.). In the process of evolution, φ grows and reaches its maximum at $N_1 = N_2$.

If we connect the quantity φ with the state probability measured by the number of complexes, entropy growth corresponds to the system evolution to the most probable state (i.e., to equilibrium).

The parameter φ is usually called the state probability, but, strictly speaking, it is not true. Actually, this parameter represents the number of dynamic states corresponding (e.g., to energy constancy). Actual probability is a ratio of this number to all possible dynamic states of a system.

Boltzmann understood that irreversible entropy growth could be considered as a manifestation of increasing molecular chaos, while the distribution asymmetry leads to a decrease in the number of complexes φ. He identified entropy to the number of complexes by the formula:

$$H = k \log \varphi \tag{1.15}$$

where k is the absolute gas constant determining the mean kinetic energy of a molecule.

Irrespective of the initial distribution, its evolution leads, in the long run, to a uniform distribution ($N_1 = N_2$).

The magnitude of k was obtained numerically for maintaining a common dimension with Clausius' entropy from the relation:

$$k = \frac{R}{N_A}$$

where R is gas constant for a gram-molecule, and N_A is the Avogadro number. Such an ensemble of systems can be represented by a distribution function expressing the probability P_f of the fact that the system chosen from the ensemble is in some concrete state.

After Boltzmann, Plank formulated principal relations between the distribution function P_f and thermodynamic properties of the macrostate in a more general form:

$$H = -k\sum P_f \ln P_f \tag{1.16}$$

for $\sum P_f = 1$. The greater φ, the greater $k \ln \varphi$. Hence, this relation satisfies the notions of entropy behaviour. It allows a more precise measurement of the disorder, which represents, in this sense, the absence of information about the exact state of the system. It is noteworthy that for systems comprising a large number of particles, all states differing from a uniform distribution are unlikely.

Collision equations derived by Boltzmann look rather natural. Therefore, their extremely paradoxical implications, namely the fact that thermodynamic entropy can only grow, seem surprising.

This caused heated debates and violent aversion to this theory. Objections of the theory opponents seemed quite well grounded. In fact, equations of mechanics are reversible in time and invariant with respect to $t \to -t$ transformation. At the same time, Boltzmann's theory determines a certain direction of time. This objection was formulated in 1876 by Loschmidt in the form of reversibility paradox. In brief, it was reduced to the following: if we examine a gas system, which is non-equilibrium in a certain state S_1, after a certain time t, this system passes into the state S_2 with $H_2 \geq H_1$. If we reverse the velocity directions of all the molecules, after the time t, the state S_2 should recover to S_1, and the entropy should be H_1 again. Meanwhile, according to the Boltzmann theorem, it should grow in this case, too. One of the proofs of Boltzmann's rightfulness was suggested by the Ehrenfest spouses after his death.

In general, abstract discussions of physics without references to the reality it tries to explain is useless and misleading and, rather often, even adverse. Thus, in the described experiment with changing directions of all velocities, something was not taken into account, namely, initial conditions. When Boltzmann's opponents applied laws of mechanics for reversing the velocities of molecules, they assumed that the system is isolated from the surroundings. However, it is unrealistic, because the isolating system also consists of atoms and molecules, which, in turn, interact both with the system and the surroundings. Hence, a reversal within a system is impossible even abstractedly – it requires a reversed motion of atoms and molecules in the entire universe, which is absurd. The Ehrenfests suggested a model that is an excellent introduction to the statistical mechanics of irreversible processes. Its simplified idea is as follows: R balls enumerated from 1 to R are placed in one box, another box being initially empty. Then a number between 1 and R is chosen at random, and the respective ball is shifted into the second box. After shifting the balls for a sufficiently long time (it can be readily simulated on a computer), the difference between the amounts of balls in these boxes tends to zero with some non-periodic fluctuations in the vicinity of zero. The probability for all the balls to gather again in one box at high enough R values is extremely low. This is the implication of thermodynamic irreversibility and entropy growth law.

When speaking about the most probable behaviour of a system, one should keep in mind that, in fact, the probability of a transition into a state with higher entropy is overwhelmingly high in comparison with a probability of its noticeable decrease, and the latter is actually never observed in nature, except for small fluctuations.

A relevant question is how and why it so happened that our Universe started from a very low entropy, which is growing ever since. The discussion of this problem leads to the Big Bang theory, which is identical to the creation act; but this is a different question misleading us away from the main topic.

In mid twentieth century, the Boltzmann ratio led Shannon to a connection between entropy and information. It was an unexpected discovery, although even the notion of thermodynamic entropy was based on certain aspects of informational approach.

Entropy is interpreted as a measure of disorder of a complicated system. Hence, the higher the entropy, the less we know about a system.

In his basic works, Shannon formulated informational entropy in the form of a relation similar to Boltzmann's formula:

$$H = -k \log P \tag{1.17}$$

where k is the proportionality factor, and P is an event probability. Some authors see a connection between the information theory and thermodynamics in this relation, assuming that the factor k in this equation is equal to the coefficient k in Boltzmann's formula Eq. (1.14). In this case, we obtain that the dimension of thermodynamic and informational entropy is the same. This is substantiated by a seemingly incontestable fact that information production, conversion, transmission, and receipt always need energy consumption, which can differ in different conditions. Therefore, instead of an actual amount of the consumed energy, the corresponding entropy is used because information is measured in entropy units. Apparently, it is caused by the understanding of the fact that transmitted information is in no way connected (e.g., with the surrounding temperature). It is unlikely that this parameter reflects the work consumed for the information transmission. This entropy is closer to a parameter evaluating dissimilarity of compositions than to thermodynamic entropy. After all, in practical applications, informational entropy becomes dimensionless and is expressed in bits – units that are in no way connected to energy.

It is assumed in thermodynamics that entropy is a measure of energy consumed for returning a system into its initial state at an irreversible change in the system.

It has been noted that Clausius formulated the notion of entropy using macroscopic parameters of thermodynamics only, whereas Boltzmann defined it using microscopic parameters only. There exists at least one region where these two aspects of entropy overlap – the theory of gas mixing and separation. In its modern form, it leads to Gibbs' paradox, which is physically grounded, and to some other paradoxes that are not physically grounded and call for investigation. Besides, it is necessary to coordinate the notion of Boltzmann's entropy, which uses as a parameter the number φ of complexes in an ensemble representing a certain number raised to the power of N ($N \approx 10^{23}$) even for a mole of gas, and informational entropy, where P is the system probability, which is always equal or less than unity.

Here, we outline some problems connected with entropy to be solved for generating a similar notion for two-phase flows.

3. Problematic aspects of entropy

There remains one unclear question connected to entropy dimension. It is well grounded that while forming the macroscopic entropy, Clausius ascribed to it a dimension equal to energy divided by temperature according to Eq. (1.1), because he could do nothing else. However, many people perceive entropy as energy, as mentioned above. Recall once more that entropy according to Boltzmann is:

$$H = k \log \varphi \quad \text{and} \quad k = \frac{R}{N_A}$$

where R is the absolute gas constant $R = 8.31$ J/mol, and N_A is the Avogadro number $N_A = 6.023 \cdot 10^{23} \frac{1}{\text{mol}}$. Hence, the Boltzmann's constant dimension gives, in essence, the dimension of the absolute gas constant. This reduces Boltzmann's entropy to Clausius' entropy, and, from the standpoint of physics, these parameters are identical. However, there is a latent drawback in this dimension, which often manifests itself in a paradoxical way.

We can illustrate this by two examples connected with mixing and, especially, separation of gases. We conceive two isolated identical volumes with different numbers of gas particles (N_1, N_2) that are different in these two volumes, but identical within each volume (Figure 1.1). Following Lorentz, we can assume that in each volume the entropy equals zero, because the probability of the presence of particles in each volume $P = 1$ and $\log 1 = 0$.

The number of complexes in a mixture is, as follows:

$$\varphi = \frac{(N_1 + N_2)!}{N_1!N_2!} \tag{1.18}$$

Resolving this expression using Stirling's formula:

$$\log N! = N \log N - N.$$

Taking this into account,

$$\begin{aligned} H &= k[(N_1 + N_2)\log(N_1 + N_2) - (N_1 + N_2) - N_1 \log N_1 + N_1 - N_2 \log N_2 + N_2] \\ &= k\{N_1[\log(N_1 + N_2) - \log N_1] + N_2[\log(N_1 + N_2) - \log N_2]\}. \end{aligned}$$

If the number of mixed particles is the same,

$$\Delta N = 2kN \log 2 \tag{1.19}$$

then this is the value of entropy increment.

The obtained result can be checked from a different standpoint. For example, Prigogine believes that, in the initial state, each of the gas volumes possesses non-zero entropy. Let each gas occupy the same volume V in a chamber separated by a partition, with the number of moles of each gas (i.e., the number of particles) being the same. After a period of time required for temperature equilibration in both volumes, the partition should be removed and the gases will start spontaneously mixing. Before mixing, the total entropy of both parts of gas, according to Prigogine, is:

$$H_1 = N\left[S_1 + C_{v_1} \ln T + R \ln\frac{V}{N}\right] + N\left[S_2 + C_{v_2} \ln T + R \ln\frac{V}{N}\right].$$

After mixing, an analogous relation acquires the form:

$$H_2 = N\left[S_1 + C_{v_1} \ln T + R \ln\frac{2V}{N}\right] + N\left[S_2 + C_{v_2} \ln T + R \ln\frac{2V}{N}\right]$$

where S_1 and S_2 are constant for each gas, C_{v_1} and C_{v_2} are heat capacities of each gas, T = absolute temperature, N = the number of particles of each gas, and R = absolute gas constant.

Entropy increment during mixing is:

$$\Delta H = 2NR \ln\frac{2V}{N} - 2NR \ln\frac{V}{N} = 2NR\left(\ln\frac{2V}{N} - \ln\frac{V}{N}\right) = 2NR\left(\ln\frac{2V \cdot N}{NV}\right) \\ = 2NR \ln 2. \tag{1.20}$$

We ignore the difference between Eqs (1.19) and (1.20), which differ in a constant. This difference is understandable and not very essential. We concentrate on the fact that mixing of gases leads to entropy growth. According to Clausius' relation, at a constant temperature this should lead to the growth of the combined system energy, which results in a paradox – violation of the energy conservation law.

Another interesting feature of the entropy of a mixture is that entropy increment of two equal quantities of different gases during their mixing amounts to:

$$\Delta H = 2N \log 2. \tag{1.21}$$

This equation is valid when the particles of these two gases differ in size or mass. They can be even identical and differ in something else (e.g., colour or charge). In this case, Eq. (1.21) is also valid for their mixing. If, however, the particles are completely identical in all parameters, entropy does not increase at their mixing. In this case, $(N_1 + N_2)$ particles just occupy the volume $2V$, and entropy increment drops to zero.

Doubts as to the connection between entropy and energy in certain physical phenomena can be illustrated by the following example. Ackeret describes the separation of gaseous system offering interesting data. To separate a mixture of gases when producing a nuclear fuel consisting of 99.3% $^{238}UF_6$ and 0.7% $^{235}UF_6$, according to computations on the basis of the mixture entropy, the energy of 0.023 *kvt/h* is needed per 1 kg of the component. However, actual energy consumption reaches 1.2 10^6 *kvt/h* (i.e., approximately 50 million times more). Proceeding from these data, Chambodal comes to a panic conclusion about the imperfection of the separation theory.

In our opinion, the point is that entropy is interpreted incorrectly, and not that the theory is imperfect. Chambodal considers entropy of a mixture of gases as energy according to Boltzmann's interpretation, assuming that it is the energy necessary and sufficient for gases separation. In fact, here, entropy features the composition heterogeneity. It is absolutely baseless to interpret entropy as a certain amount of energy required for the realisation of the respective state of a system. In such processes, entropy can only give an estimate of the efficiency of the components separation (i.e., of the extent of the process completion). This is discussed in more detail in Chapter 5. In this connection, we make an attempt to define the physics of entropy more precisely.

According to the already mentioned empirical rule of Dulong and Petit for atomic gases, first, their heat capacity is independent of temperature, and, second, the specific heat capacity of a mole of gas is constant and equals approximately:

$$C_p \approx 21\ \mathrm{J/molK}. \tag{1.22}$$

This rule expresses the energy conservation law, because a mole of different gases contains the same number of particles having individual masses for each gas. Fine particles gather higher speeds than heavy ones. The temperature of all these gases rises by the same value at the same energy consumption.

We examine the specific heat capacity parameter, whose dimension is analogous to that of thermodynamic entropy:

$$C_p = \frac{dQ}{dT} \tag{1.23}$$

where C_p is heat capacity of gas at a constant pressure. According to Eq. (1.22), the specific heat capacity per unit mass is expressed as:

$$C_p = \frac{21}{\mu} \text{J/K}$$

where μ is the mass of a mole of the corresponding gas.

This hyperbolic relation shows that for light gases this parameter is of greater importance than for heavy ones. Hence, temperature increase of a unit mass of light gases by one degree requires greater heat consumption than that of heavy gases, which seems paradoxical.

Now we can revert to the entropy expression of Clausius:

$$dH = \frac{dQ}{T},$$

whence and from Eq. (1.23), two equalities follow:

$$dQ = TdH \text{ and}$$

$$dQ = C_p dT,$$

hence,

$$TdH = C_p dT.$$

We separate variables and take the integral

$$\int_{T_1}^{T_2} \frac{dT}{T} = \frac{1}{c_p} \int_{T_1}^{T_2} dH.$$

We obtain

$$\Delta H = C_p \log \frac{T_2}{T_1}$$

where ΔH is entropy increment with temperature change from T_1 to T_2.

Temperature ratio adds nothing to the dimension of C_p. Logarithm in the temperature ratio considerably levels the effect of their difference on the entropy increment value, which is always positive. This shows that entropy grows at temperature increase from T_1 to T_2, but its absolute value is mostly determined by the specific heat capacity of gas.

All these paradoxes can be overcome using one idea only. It consists in the assumption that entropy H is dimensionless not only in the theory of information, but also in thermodynamics. In fact, it is actually true. Following Clausius' relation

$$dH = \frac{dQ}{T},$$

entropy is interpreted as a parameter connected to energy.

Here, we come across a strange logical situation. In this case, the temperature T becomes a dimensionless parameter with respect to energy and represents a kind of a dimensionless coefficient devoid of physical meaning.

The fact that temperature is a carrier of energy is clear even at an intuitive level and laws of physics point to that, too. Recall Gay-Lussac's law stating that the energy of an ideal gas depends on temperature only, being independent of volume and pressure. We can also mention phenomena arising at the thermal expansion of solids. They are determined by temperature only and develop huge forces.

It is perfectly clear that the definition of temperature must be based on a physical quantity characterising the state of a body, which is automatically the same for any various bodies that are in thermal equilibrium with each other.

It has been known since long ago that the mean kinetic energy of translational motion of particles (atoms or molecules) of a body possesses this remarkable property. Because of this, the mean kinetic energy of the translational motion of particles in a system or a body is chosen as a measure of temperature. As early as in the nineteenth century, it was known from fundamental physics that temperature depends on the kinetic energy of translational motion of molecules and is defined by the relation:

$$T = \left\langle \frac{1}{3} m v^2 \right\rangle \tag{1.24}$$

where m is the particle mass, $v =$ its mean velocity. Here, the mean value can imply either a value averaged over the velocities of the body particles at the same moment of time, or a value averaged over the velocities of one and the same particle at different moments of time. Both these definitions are absolutely equivalent. According to Eq. (1.24), the dimension of temperature is that of energy, and it can be measured in energy units, say, in ergs. However, because of two reasons, this parameter is extremely inconvenient.

First, the energy of thermal motion of particles is negligibly small in comparison with erg. Second, a direct measurement of temperature as particles energy is extremely difficult. No one would think of measuring temperature in this way having a possibility of using degrees – a simple and practical measurement unit for this parameter. It should be emphasised that a degree of temperature also has the dimension of energy. A conversion factor has been found showing the fraction of erg contained in one degree of temperature. This factor corresponds to the Boltzmann constant and equals:

$$k = 1.38 \cdot 10^{-16}\ \mathrm{erg/K}.$$

Total kinetic energy of particles in 1 g-molecule of a substance per one degree equals the following:

$$k \cdot N_A = 1.38 \cdot 10^{-16} \cdot 6.02 \cdot 10^{23}\mathrm{erg} = 8.31\ \mathrm{J}.$$

Now it is clear that if temperature is expressed in degrees, its magnitude in ergs equals kT, and the relation of Eq. (1.24) should be written as:

$$kT = \left\langle \frac{1}{3} m v^2 \right\rangle. \quad (1.25)$$

Here, we have to emphasise the fact that the dimension of energy refers to the temperature parameter, and not to the Boltzmann constant.

Hence, it is clear that entropy in the Clausius' formula is a dimensionless parameter, just as the informational entropy.

From this standpoint, a specific heat capacity is also dimensionless and most probably characterises thermal mobility of a specific system. This fact removes all absurdities mentioned in this chapter. Obviously, all misunderstandings arose because the dimension erg/°K was assigned to the Boltzmann factor (k). This dimension is perceived as energy, and, in most cases, this factor is interpreted exactly in this way. In fact, however, it is dimensionless, because both erg and degree refer to energy, and, thus, the dimension of this factor is zero. The only case – a combination of this factor with temperature kT – actually expresses energy. We recall that this factor constitutes a ratio between the temperature expressed in degrees and energy expressed in ergs. As for Boltzmann's entropy:

$$H = k \log \varphi,$$

in this relation, k is dimensionless. Hence, the entropy according to Clausius and Boltzmann is dimensionless. This makes it possible to compare and combine it with entropy of other processes that are not connected to heat transfer, for example, informational entropy, etc.

There is another aspect of this problem that prevents further discussion. By way of example, we revert to the gaseous system in Figure 1.1. Here, the entropy value is determined through the number of complexes defined by the following expression:

$$\varphi = \frac{(N_1 + N_2)!}{N_1!N_2!} = \frac{N!}{N_1!N_2!} \quad (1.26)$$

the entropy being equal to:

$$H = k \ln \varphi. \quad (1.27)$$

The number of molecules in gas is enormous. One gram-molecule contains $N_A \approx 6 \cdot 10^{23}$ of them. Multiplying or dividing Eq. (1.27) by N_A, we obtain

$$H = k \ln \varphi \cdot N_A. \quad (1.28)$$

Here, the value of the factor k is not fundamental and does not affect the logics of our reasoning. Therefore, we concentrate on the expression under the logarithm sign. Because each particle can be located either in the right-hand or in the left-hand compartment (Figure 1.1), the probability of its state is $P = \frac{1}{2}$, and the number of complexes in the ensemble is approximately 2^N. This problem will be discussed in detail in the next chapters. Here, we have to underline that, in any case, φ significantly exceeds N_A. We expand the logarithm in Eq. (1.28) as follows:

$$H = k(\ln \varphi \pm \ln N_A) \quad (1.29)$$

and divide both parts of Eq. (1.29) by N_A

$$\frac{H}{N_A} = \frac{k \ln \varphi}{N_A} \pm \frac{k \ln N_A}{N_A}.$$

The second summand in the right-hand part of this equation is vanishingly small. Hence, it does not affect entropy and can be discarded. It points to another subtle property of entropy. The values of its parameters under the sign of logarithm can comprise countless numbers of complexes of the system, but they can also have the sense of probability expressed by fractions of unity. The ratio between these values appearing in the expression of entropy is of utmost importance.

In conclusion, we underline once more that the entropy of any physical system (thermodynamic, informational, compound, etc.) determining its uncertainty degree is dimensionless. Therefore, entropies of any systems can be compared and combined with each other.

CHAPTER

Invariants for Continuous Flows

2

CHAPTER OUTLINE

1. Substantiation of the necessity to model technological processes 17
2. Certain aspects of continuous flows modeling 18
3. Invariants for continuous flows 22
4. Analysis of flow parameters using similarity criteria 28

1. Substantiation of the necessity to model technological processes

Most technologies used in modern industry cannot be described analytically. They combine various processes based on numerous elements, as well as on common laws of physics, chemistry, hydrodynamics, thermodynamics, etc., which lack reliable computation methods. Therefore, each specific problem is being solved in a different way. One of them is the development of mathematical models based on physical laws and mathematical analytical methods. However, the analysis of such models can be realized rather seldom, for the simplest cases only. Meanwhile, the practice needs particular solutions.

Industrial processes are characterized by numerous peculiar features and internal bonds. Most often, it is impossible to consider them thoroughly and precisely, and, therefore, the most common ones are singled out for the scientific analysis. This leads to a rather approximate model of the process, which makes use of mathematical tools and knowledge derived from many fields of physics, hydraulics, mineralogy, etc. When developing a mathematical model, one has to put aside many hard-to-account-for features of flows. As a result, the model is only relatively related to the process or object. However, any relativity is recognizable, because it possesses characteristic features of an actual object.

Hence, the results of mathematical study of a problem and respective design solutions should not be accepted as solely possible or strictly conditioned. It is not always possible to single out the optimal choice, and, thus, one has to choose by comparing the advantages and disadvantages of the obtained possibilities.

A scientific analysis is based on quantitative connections between various factors of a process. Experimental data obtained both on industrial equipment and on laboratory experimental facilities are generalized, and empirical relations are derived. They are, as a rule, of a particular character, and their application beyond the range of parameters where they were obtained often leads to gross errors. It is not always possible to expand the range of parameters in the experiment. For example, one cannot

Entropic Invariants of Two-Phase Flows. http://dx.doi.org/10.1016/B978-0-12-801458-5.00002-2

study a natural-sized facility under laboratory conditions. Therefore, it is usually rather hard to proceed from a laboratory model to a pilot plant. Such a transition involves numerous mistakes and calls for their corrections.

The variety of bonds between the properties of an object and process parameters complicated by insufficient insight into the physics of the phenomena leads to intricate calculated relations and makes it difficult to establish quantitative regularities.

A transition to generalized variables made up from elementary factors of the process according to certain rules makes it somewhat easier to overcome this difficulty. These new variables are dimensionless and have a certain physical meaning. They allow establishing relations between generalizing complexes combining the factors of the process, and not between these numerous factors.

These methods are most successfully applicable to single-phase flows both in pipes and in apparatuses of various shapes.

We emphasize that the combination of factors is not a formal procedure. In real processes, the effect of separate factors is additive, and not individual. Therefore, if these factors are combined into a single complex, the latter reflects the total state of the process.

The use of generalized variables, including several factors each, makes the process description more general, because a single value of a complex can be realized, strictly speaking, at an infinitely large number of combinations of numerical values of factors it contains. Hence, these complexes can characterize not only individual phenomena or processes, but also a group of similar phenomena or processes, for which such complexes have the same numerical values. This forms the basis of the notion of physical similarity. Therefore, complexes of such kind are called similarity criteria. They are applied in modeling complicated processes and apparatuses, experimental data processing, and analytical solution of technological problems.

In addition, it has turned out that intricate differential equations derived by analytical methods are valuable themselves. Their exact solution is practically impossible. However, on the basis of these equations, without solving them, one can formulate correctly similarity criteria, set an experiment competently, and process the obtained results.

2. Certain aspects of continuous flows modeling

A continuous medium implies either a condensed liquid or an elastic gas. If the gas pressure is insignificant, the change in its volume can be neglected, and the gas can be considered as an elastic medium. Processes in moving continuous media have been studied since very long ago. The first works on liquid flows in pipes were published by Hagen as early as 1839. However, systematic study of these phenomena became possible only after Reynolds' remarkable works had been published.

Main invariants for moving systems were based on elementarily simple models. Thus, principal hydraulic equations were derived from balance ratios. Here, such important features of flows as flow structure, turbulence with its developed spectrum of fluctuations, and situation on the channel walls, were not taken into account, whereas the respective balances were accounted for. Even such a simplified model can help define useful regularities and similarity criteria.

We examine this in brief, because we intend to use similar methods for the study of two-phase flows. First, we analyze the kinetic aspect of the problem. We place a motionless infinitesimal elementary cuboid with edges d_x; d_y; d_z into a flow (Figure 2.1) and examine flows through pairs of

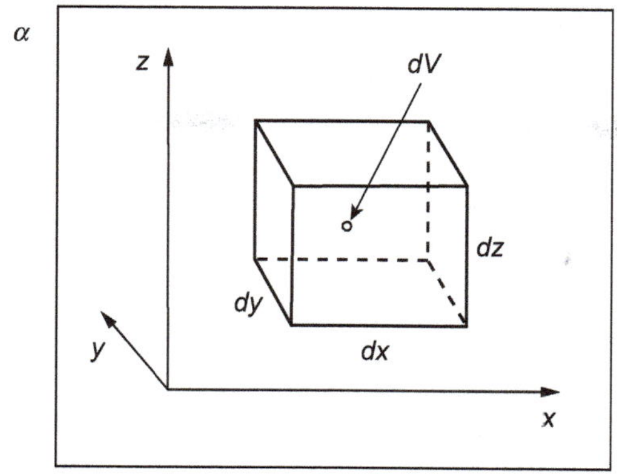

FIGURE 2.1 Elementary volume of the movable medium.

parallel faces $dxdy$; $dxdz$; $dydz$ during an infinitesimal time interval dt, as well as changes in a selected elementary volume $dV = dxdydz$.

An elementary amount of moving medium entering a motionless contour along x axis through $dydz$ face at the velocity w_x during the time dt equals:

$$dG_x^+ = \rho w_x dydzdt,$$

while the amount:

$$dG_x^- = \left[\rho w_x + \frac{\partial(\rho w_x)}{\partial x}\right] dydzdt, \tag{2.1}$$

goes out through a parallel face. Here, ρ is the medium density in a point of the contour, $w_x =$ the velocity along x-axis in a point of the contour. After opening the brackets and reductions, the difference between dG_x^+ and dG_x^- for x-axis is:

$$-\frac{\partial(\rho w_x)}{\partial x} dxdydzdt = -\frac{\partial(\rho w_x)}{\partial x} dVdt. \tag{2.2}$$

Analogous differences for y- and z-axes can be written as:

$$-\frac{\partial(\rho w_y)}{\partial y} dVdt \quad \text{and} \quad -\frac{\partial(\rho w_z)}{\partial z} dVdt.$$

The mass contained in the volume dV at the moment t equals ρdV. Because ρ can vary along the coordinates (x; y; z) and in time (t), mass change in this volume by the moment $t + dt$ can be written in partial derivatives:

$$\frac{\partial(\rho dV)}{\partial t} dt = \frac{\partial \rho}{\partial t} dVdt. \tag{2.3}$$

Now, we gather all found elements of balance and, after reducing by dV and dt, obtain:

$$-\frac{\partial \rho}{\partial t} = \frac{\partial(\rho w_x)}{\partial x} + \frac{\partial(\rho w_y)}{\partial y} + \frac{\partial(\rho w_z)}{\partial z}.$$

This can be written in a different form:

$$\frac{\partial \rho}{\partial t}+\frac{\partial (\rho w_x)}{\partial x}+\frac{\partial (\rho w_y)}{\partial y}+\frac{\partial (\rho w_z)}{\partial z}=0. \tag{2.4}$$

which is called a flow continuity equation. Integration of this equation with concrete single-valued conditions leads to the mass conservation law in an integral form.

Now we examine the dynamic aspect of the problem. A cardinal problem in the analysis of the momentum transfer is to determine pressure (p) and velocity (w) in a certain point of the flow at an arbitrary moment of time t:

$$p=p(x;y;z;t); \quad w=w(x;y;z;t).$$

We examine an elementary cuboid again. First, we analyze the situation as applied to one coordinate axis (x). The obtained results can be extended to other axes (y, z). We consider forces of different nature one after another:

1. along x-axis on the left-hand face $dydz$ equal $pdydz$. The force acting on a parallel face from the right is $\left(p+\frac{\partial p}{\partial x}dx\right)dydz$. The difference between these normal forces is:

$$pdydz-\left(p+\frac{\partial p}{\partial x}dx\right)dydz=-\frac{\partial p}{\partial x}dxdydz=-\frac{\partial p}{\partial x}dV; \tag{2.5}$$

2. Tangential forces (of internal friction, shear, viscosity $-\tau$). The force $\tau_1 dxdy$ acts along the lower face, and for the upper face, we can write $\tau_2 dxdy$. We denote the velocity at the lower face by w_x, and at the transition to the upper face along z-axis, the velocity acquires a certain value $w_x+\frac{\partial w_x}{\partial z}dz$.
   According to the Newton law, the relation between the tangential force and velocity gradient in a liquid or gas is linear:

$$\tau=-\mu\frac{dw}{dz}, \tag{2.6}$$

   where μ is the dynamic viscosity factor of the medium. Taking this into account, we can write for the upper and lower faces:

$$\tau_1 dxdy=-\mu\frac{\partial w_x}{\partial z}dxdy; \tag{2.7}$$

$$\tau_2 dxdy=-\mu\frac{\partial}{\partial z}\left(w_x+\frac{\partial w_x}{\partial z}dz\right)dxdy. \tag{2.8}$$

   After opening brackets and reductions, the difference between these forces is:

$$(\tau_1-\tau_2)dxdy=\mu\frac{\partial^2 w_x}{\partial z^2}dxdydz=\mu\frac{\partial^2 w_x}{\partial z^2}dV. \tag{2.9}$$

Similar tangential forces along x-axis act on other pairs of faces, and by analogy with previously obtained, we can write:

$$\mu\frac{\partial^2 w_x}{\partial y^2}dV \quad \text{and} \quad \mu\frac{\partial w_x^2}{\partial x}dV.$$

The total balance of all tangential forces along x-axis is expressed by the sum:

$$\mu\left(\frac{\partial^2 w_x}{\partial x_2}+\frac{\partial^2 w_x}{\partial y^2}+\frac{\partial w_x}{\partial z^2}\right)dV=\mu\nabla^2 w_x dV, \tag{2.10}$$

where ∇^2 is the Laplace operator (Laplacian).

3. External mass forces
External mass forces act on the mass of liquid or gas with the density ρ in the volume dV. We denote by P_x the projection of the resultant unit mass force on x-axis. Then the total force acting on the volume under study is $P_x\rho dV$.
4. According to the Newton law, the inertia force acting on this volume equals the product of its mass on the acceleration along x-axis:

$$\rho dV\frac{dw_x}{dt}.$$

Making a net balance of these four constituents after the reduction by dV with respect to the axis x, we can write:

$$-\frac{\partial p}{\partial x}+\mu\nabla^2 w_x+P_x\rho=\rho\frac{dw_x}{dt}. \tag{2.11}$$

Dividing each constituent by ρ and substituting $\frac{\mu}{\rho}=\nu$ (kinematic viscosity coefficient), we can finally write:

$$\frac{1}{\rho}\frac{\partial p}{\partial x}=P_x-\frac{\partial w_x}{d\tau}+\nu\nabla^2 w_x. \tag{2.12}$$

If we write, by analogy, such expressions for y and z axes, the obtained system of equations is called Navier–Stokes equation.

Note that during the derivation of equations in this Chapter, we are not interested in the signs of differences at the stage of intermediate relationships.

Following mathematical rules, all these increments were assumed to be positive, and the sign should appear at the definition of integration limits or single-valued condition (e.g., boundary conditions) or concrete definition of the process conditions (for example, gravity force is always opposite to an ascending flow).

It is assumed that the system of Navier–Stokes equations together with the continuity equation characterizes the motion of a medium. To obtain a single-valued solution of the system, even for the simplest conditions of a hollow pipe, one needs to specify initial values of velocity fields in time and space taking into account that the velocity must vanish on the channel walls and on the surface of all solids submerged into the flow.

The solution of this system of equations was discussed by many researchers. As early as the 1920s, Keller and Friedman showed that to determine statistical moments of any order for hydrodynamic fields of single-phase turbulent flows, an infinite system of equation should be solved, because this system is not closed. It becomes possible to solve this system by introducing various assumptions, which idealize the moving medium. We know idealized flows of Newton, Euler, Couette, Poiseuille, Haden, and others. These solutions are of a certain theoretical interest, but no practical importance in the general case.

Thus, it is impossible to find an exact solution of the obtained system of equations. However, it contains information about the flow character, which can be extracted in the form of dimensionless complex parameters. It is possible to determine these parameters using the methods of similarity theory.

3. Invariants for continuous flows

It is generally accepted that groups of resembling processes, phenomena, or objects, which can be mathematically described using similarity criteria or invariants, are similar. The notion of geometrical similarity is taught at a secondary school. Physical similarity is different. Two physical phenomena can be considered similar if their homonymous characteristics in analogous points of geometrically similar systems differ only by a coefficient, which is constant in all the points. A mathematical description of such systems is identical. We examine two geometrically similar devices (Figure 2.2) and call the bigger one an apparatus (A) and the smaller one a model (B). Imagine movable media with differing properties (density, viscosity, heat capacity, etc.) flowing through these devices.

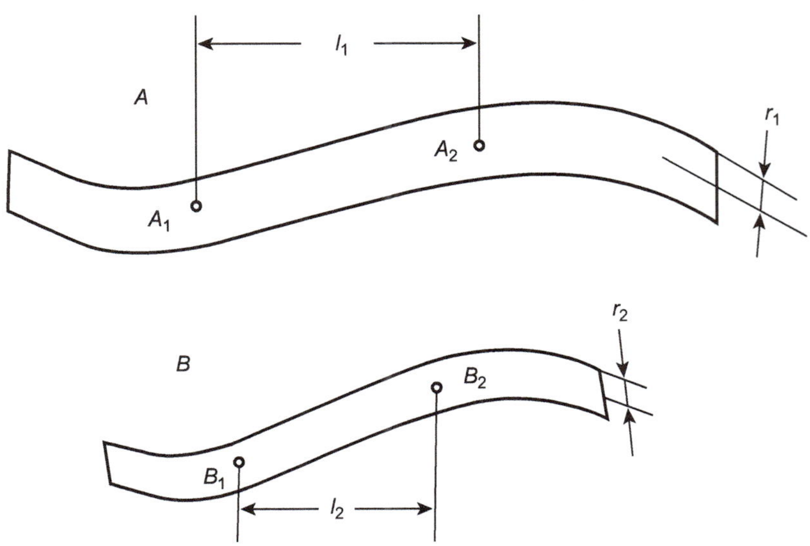

FIGURE 2.2 Geometrically similar channels.

We choose two pairs of similarly located points A_1 and A_2; B_1 and B_2 in these systems. Geometrical parameters of each pair are characterized by a certain correspondence of the following type:

$$\frac{x_1}{D_1} = \frac{x_2}{D_2} = \cdots$$

The ratio:

$$\frac{x_1}{x_2} = \frac{D_1}{D_2} = \cdots = m_1 \tag{2.13}$$

where m_1 is a geometrical similarity factor, which is constant for each pair of similarly located points in the devices under study.

By the definition of physical similarity for any pair of similar points A_1 and B_1; A_2 and B_2 and any other pairs, the following equalities should hold:

$$\frac{\rho_1}{\rho_2} = m_\rho; \quad \frac{\mu_1}{\mu_2} = m_\mu; \quad \frac{w_1}{w_2} = m_w; \quad \frac{c_1}{c_2} = m_c \text{ etc.} \tag{2.14}$$

In these expressions, m_i are similarity factors. Naturally, they are different for different parameters $(m_\rho \neq m_w \neq m_c \neq \cdots)$, but within the limits of one parameter, they are constant for each pair of similarly located points. Each similarity factor is a sort of scale for the characteristic of the respective physical quantity.

Similarity criteria can be derived on the basis of the notion of similarity factors. We show below the derivation of one of such criteria from the flow continuity equation. We can write the following for two geometrically similar flows:

$$\frac{\partial \rho_1}{\partial t_1} = \frac{\partial(\rho_1 w_1)}{\partial x_1} + \cdots = 0; \tag{2.15}$$

$$\frac{\partial \rho_2}{\partial t_2} = \frac{\partial(\rho_2 w_2)}{\partial x_2} + \cdots = 0. \tag{2.16}$$

The rest of the summands are omitted, because their structure coincides with that of the second summands. We apply scaling transformation to these two equations:

$$\rho_1 = m_\rho \rho_2; \quad t_1 = m_t t_2; \quad w_1 = m_w w_2; \quad x_1 = m_l x_2. \tag{2.17}$$

Geometrical similarity factor m_i is constant for all the coordinates (x; y; z), as well as for linear dimensions l_1 and l_2.

Now, we substitute the values of all parameters of the process in the apparatus into Eq. (2.15) expressing them through respective similarity factors and process parameters in the model. By factoring constant multipliers out of the derivatives, we obtain a new expression for the apparatus through the parameters of the model.

$$\frac{m_\rho}{m_t}\frac{\partial \rho_2}{\partial t_2} + \frac{m_\rho m_w}{m_l}\frac{\partial(\rho_2 w_2)}{\partial x_2} + \cdots = 0. \tag{2.18}$$

Compare the obtained expression with the continuity equation for the model written above. These expressions can be adequate in the only case when the complexes comprising similarity factors can be factored out and reduced, because the right-hand part is zero. This means that these multipliers are equal:

$$\frac{m_\rho}{m_t} = \frac{m_\rho m_w}{m_l}. \tag{2.19}$$

Here, we can reduce by m_ρ obtaining, as a result:

$$\frac{1}{m_t} = \frac{m_w}{m_l} \tag{2.20}$$

However, it is known that:

$$m_t = \frac{t_1}{t_2}; \quad m_w = \frac{w_1}{w_2}; \quad m_l = \frac{l_1}{l_2}. \tag{2.21}$$

Substitute these values into the final expression:

$$\frac{t_2}{t_1} = \frac{w_1}{w_2} \cdot \frac{l_2}{l_1} \tag{2.22}$$

and collect terms with the same indices in different parts of the equality. We obtain the following:

$$\frac{w_1 t_1}{l_1} = \frac{w_2 t_2}{l_2} = \frac{wt}{l} = idem. \tag{2.23}$$

This points to the validity of the obtained result for all such flows. This complex is dimensionless. The coincidence of the numerical value of a dimensionless complex or a group of complexes is a necessary and sufficient similarity condition for several systems, objects, flows, or processes. The obtained complex is called homochronism criterion and denoted by:

$$Ho = \frac{wt}{l}. \tag{2.24}$$

Most often, such criteria are attributed to great scientists and are denoted by the two first letters of their names.

The physical meaning of homochronism criterion is clear from the analysis prerequisites, namely, two opposite effects are compared – forced transfer of the medium and its accumulation.

The growth of *Ho* value testifies that the influence of the factor in the numerator grows (i.e., mass transfer increases). With decreasing *Ho*, the role of the factor in the denominator grows (i.e., mass accumulation prevails). For stationary processes without mass accumulation in the working volume, *Ho* criterion is degenerated, and its numerical values tend to infinity.

Note another important aspect. In many cases, a complete geometrical similarity is not necessary, approximate one being sufficient. Geometrical sizes can affect a process in various ways. For example, it is clear that shelf thickness in a reactor does not affect the reaction character. Therefore, it does not require the observance of geometrical similarity, such as:

$$\frac{\delta_1}{l_1} = \frac{\delta_2}{l_2}. \tag{2.25}$$

An example of the velocity profile of a flow entering an apparatus is less evident. Its formation is completed at the initial segment of the channel, and then the velocity field does not appreciably change. Beyond the input segment, the longitudinal coordinate influence degenerates, and, in this sense, geometrical similarity becomes nonobligatory.

Using the same reasoning, we make an attempt to derive similarity criterion from the system of differential Navier–Stokes equations analyzing similar flows with respect to z axis. We can write the following for analogous points of the apparatus and the model:

$$\frac{1}{\rho_1}\frac{\partial p_1}{\partial z_1} = P_1 - \left(w_1\frac{\partial w_1}{\partial z_1} + \cdots\right) + \nu_1\left(\frac{\partial^2 w_1}{\partial z_1^2} + \cdots\right)$$
$$\frac{1}{\rho_2}\frac{\partial p_2}{\partial z_2} = P_2 - \left(w_2\frac{\partial w_2}{\partial z_2} + \cdots\right) + \nu_2\left(\frac{\partial^2 w_2}{\partial z_2^2} + \cdots\right). \qquad (2.26)$$

We also express all characteristics of the apparatus through similarity factors and characteristics of the model:

$$\rho_1 = m_\rho\rho_2; \quad p_1 = m_p p_2; \quad P_1 = m_m P_2; \quad z_1 = m_l z_2; \quad w_1 = m_w w_2; \quad \nu_1 = m_\nu \nu_2. \qquad (2.27)$$

Note that m_m is the same for all mass forces. We also substitute all these relations into the first Eq. (2.26) as follows:

$$\frac{m_p}{m_\rho m_l}\frac{1}{\rho_2}\frac{\partial p_2}{\partial z_2} = m_m P_2 - \frac{m_w^2}{m_l}\left(w_2\frac{\partial w_2}{\partial z_2} + \cdots\right) + \frac{m_\nu m_w}{m_l^2}\nu_2\left(\frac{\partial w_2}{\partial z_2^2} + \cdots\right). \qquad (2.28)$$

The meaning of the second derivative is a quotient of twofold division of the differentiated value by the argument. Therefore, m_l is written in the last expression in the second degree. Hence, for the same reasons as in the first case:

$$\frac{m_p}{m_\rho m_l} = m_m = \frac{m_w^2}{m_l} = \frac{m_\nu m_w}{m_l^2}. \qquad (2.29)$$

The physical meaning of these parts of the equality are as follows: $\frac{m_p}{m_\rho m_l}$ – pressure forces; m_m – mass forces; $\frac{m_w^2}{m_l}$ – inertia forces; $\frac{m_\nu m_w}{m_l}$ – viscosity forces. Usually, these parts of the equality are examined in pairs.

1. Comparison of pressure and inertia forces:

$$\frac{m_p}{m_\rho m_l} = \frac{m_w^2}{m_l}. \qquad (2.30)$$

After reducing by m_l and substituting multipliers:

$$m_\rho = \frac{\rho_1}{\rho_2}; \quad m_p = \frac{p_1}{p_2}; \quad m_w = \frac{w_1}{w_2}, \qquad (2.31)$$

we obtain the following:

$$\frac{p_1 \rho_2}{p_2 \rho_1} = \frac{w_1^2}{w_2^2}.$$

We separate values with the same indices:

$$\frac{p_1}{\rho_1 w_1^2} = \frac{p_2}{\rho_2 w_2^2} = \frac{p}{\rho w^2} = Eu = idem, \tag{2.32}$$

Eu being Euler's criterion. The meaning of this criterion consists in the definition of the ratio between inertia and pressure forces. Most often, in practical problems, pressure drop over a certain section is of interest, and not the absolute pressure value. Therefore, a somewhat different expression is used:

$$Eu = \frac{\Delta p}{\rho w^2}. \tag{2.33}$$

2. Comparison of mass and inertia forces. In this case:

$$m_m = \frac{m_w^2}{m_l}. \tag{2.34}$$

For vertical z axis, any mass is characterized by the acceleration g. Taking this into account, the substitution of m_w; m_l and $m_m = \frac{g_1}{g_2}$ values gives:

$$\frac{g_1}{g_2} = \frac{w_1^2 l_2}{w_2^2 l_1}.$$

Hence,

$$\frac{g_1 l_1}{w_1^2} = \frac{g_2 l_2}{w_2^2} = \frac{gl}{w^2} = Fr = idem, \tag{2.35}$$

Fr being Froude's criterion, which is a measure of the ratio between mass and inertia forces. The comparison of these forces predetermines the character of critical flow processes under study, and, therefore, this criterion plays a determining role therein, which is experimentally confirmed.

3. Comparison of inertia and viscosity forces:

$$\frac{m_w^2}{m_l} = \frac{m_\nu m_w}{m_l^2}. \tag{2.36}$$

After reducing by $\frac{m_w}{m_l}$ and substituting all similarity factors, we obtain the following:

$$\frac{w_1}{w_2} = \frac{\nu_1 l_2}{\nu_2 l_1}. \tag{2.37}$$

Hence:

$$\frac{w_1 l_1}{\nu_1} = \frac{w_2 l_2}{\nu_1} = \frac{wl}{\nu} = \text{Re} = idem. \tag{2.38}$$

In this way, we obtain the Reynolds criterion, which is widely used to describe liquid and gas flows. The channel diameter is usually accepted as a linear dimension in the Reynolds criterion:

$$\text{Re} = \frac{wd}{\nu} \tag{2.39}$$

To obtain other similarity complexes, we can use these three main ones, for example, the ratio between the gravity and viscosity forces can be written as *Fr* Re.

We can also obtain other combinations of similarity criteria having a clear physical meaning.

Usually, it is impossible to establish quantitative relations between similarity criteria in a purely theoretical way. It is realized in each specific case by a special experiment. Such relations are called criterial equations of $Eu = f(\text{Re})$ type (Figure 2.3). Such relations are valid only within experimentally checked ranges of similarity criteria variation.

Similarity criteria should not be interpreted as parameters defining the ratios of respective forces, because their values differ in various points of the flow. They should be taken as a measure

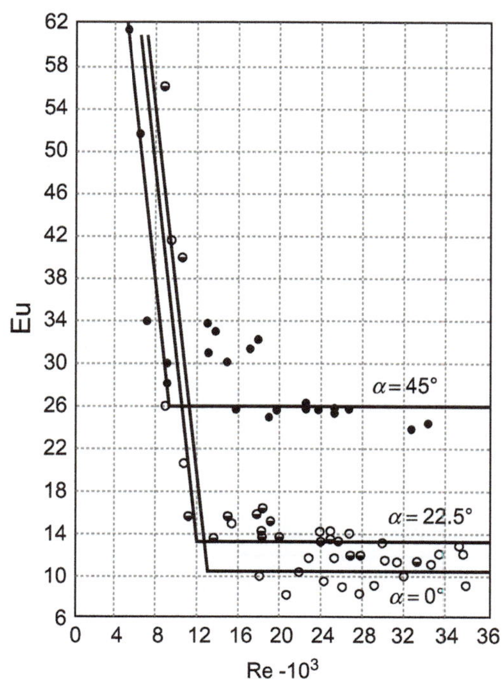

FIGURE 2.3 Determining the self-similarity region by the criterial relation $Eu = f(\text{Re})$.

characterizing correctly the ratio between the respective forces, namely, for instance, the higher Re number value, the more inertial forces exceed friction forces in specific flows.

4. Analysis of flow parameters using similarity criteria

Concentrate on the empirically found Reynolds criterion, which represents a measure of the ratio between the inertial force and the internal friction force. It is the most important characteristic of a flow, because the main properties of a flow depend on the ratio between these forces. Inertia forces lead to the approach of initially distant volumes of the medium, which promotes the formation of flow inhomogeneities. Viscosity forces, on the contrary, lead to velocity leveling in close points (i.e., to the smoothing of the inhomogeneities). At low Re values, when the viscosity forces dominate over the inertia forces, all the fields characterizing the flow change smoothly, and the flow is laminar.

Here, the paths of separate particles represent straight lines parallel to the flow axis, which form a system of nonintersecting curves at the turns. Because of the internal friction, the velocity of elementary jets in various points of the cross-section is different. The layer adjacent to the pipe wall experiences a much higher friction than the friction between liquid layers. This near-wall layer remains practically motionless. The velocity of its translational motion is zero. The next (toward the flow axis) layer is braked by friction against the near-wall layer (i.e., much more weakly). This second layer moves in the direction of the entire flow at a higher velocity. All subsequent liquid layers move parallel to each other, their velocities increasing up to the maximum value on the pipe axis (Figure 2.4).

Velocity distribution over the cross-section of a circular pipe with the radius R can be established by equating the internal friction forces between layers moving at various velocities to hydrostatic pressure forces over the length of the segment under study (Figure 2.5).

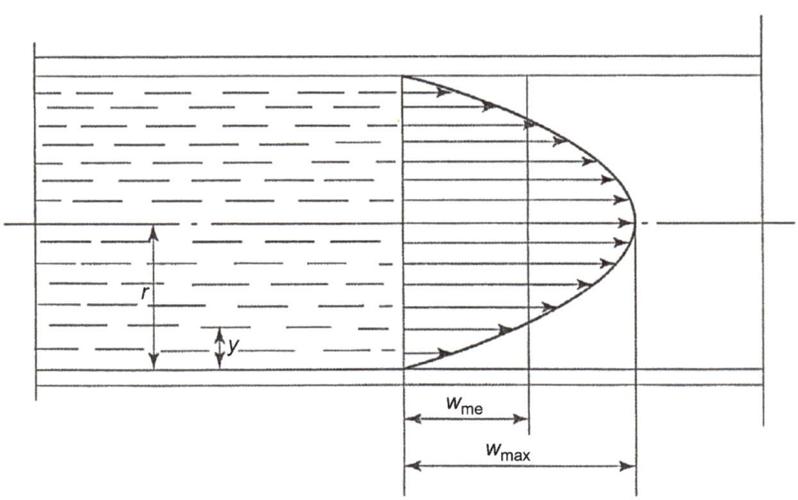

FIGURE 2.4 Flow velocity distribution in a round tube at a laminar flow regime.

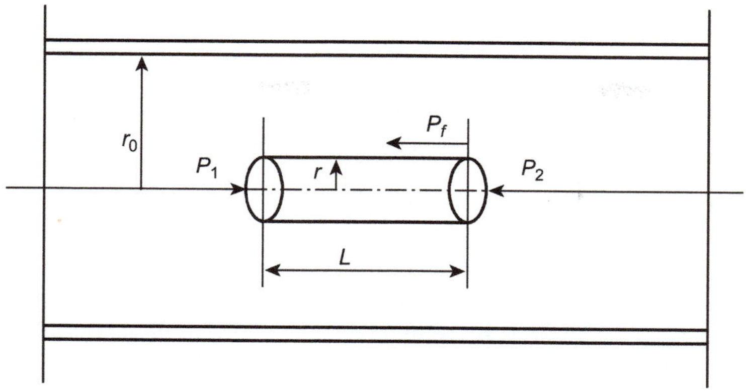

FIGURE 2.5 Dynamic equilibrium of a flow element.

We assume that the cylinder moves from left to right. The resulting hydrostatic pressure force acting on it equals:

$$P = (P_1 - P_2)\pi r^2.$$

According to the Newton law, the friction force along the cylinder generatrix is:

$$P_{fr} = \mu F \frac{dw}{dt} = -\mu 2\pi r l \frac{dw}{dr}$$

For a steady motion,

$$dw = -\frac{P_1 - P_2}{2\mu l} r dr.$$

After integration, we obtain:

$$w = \frac{P_1 - P_2}{4\mu l}\left(r_0^2 - r^2\right). \tag{2.40}$$

Equation (2.40) shows that velocity distribution in a laminar flow obeys a parabolic law. The velocity is maximum on the pipeline axis at $z = 0$ and equals:

$$w_{\max} = \frac{P_1 - P_2}{4\mu l} r_0^2. \tag{2.41}$$

Elementary volumetric rate through a ring dr thick is:

$$dV = wdF = \frac{P_1 - P_2}{4\mu l}\left(r_0^2 - r^2\right) 2\pi r dr.$$

Hence, if we integrate over the radius from the flow axis to the wall, the total medium flow rate is:

$$V = \frac{\pi(P_1 - P_2)}{8\mu l} r_0^4. \tag{2.42}$$

From this equation, we can find the mean flow velocity:

$$w_m = \frac{(P_1 - P_2)}{8\mu l} r_0^2. \tag{2.43}$$

It follows from the comparison of Eqs (2.41) and (2.43) that the mean velocity of a laminar flow is twice smaller than the maximum (axial) one:

$$w_m = 0.5 w_{\max}.$$

At high Re values, the smoothing action of the viscosity forces weakens, which leads to the formation of local inhomogeneities in a flow (i.e., the flow becomes turbulent). The main reason of turbulence origination in moving liquids or gases is the hydrodynamic stability loss. Turbulent motion is characterized by chaotic velocity fluctuations in time and space in both axial and transverse directions. Similarly, pressure in various points of the flow also changes. This phenomenon is considerably complicated by the chaotic pulsating stirring of local volumes of the medium. This generates a specific phenomenon – turbulent diffusion, whose intensity exceeds regular molecular diffusion by many orders. All this predetermines changes in the resistance laws in this region of the flow, which are accompanied by a drastic growth of hydrodynamic losses (square law). Owing to this, leveling of the mean velocity profile in the flow core and its abrupt drop in the near-wall region occur (Figure 2.6). Thus, the flow regime is characterized by the Reynolds number value.

This predetermines a simple relation between flow properties and the value of this criterion. A steady laminar flow corresponds to low Re values. With increasing Re, the stability of such motion decreases because of a relative growth of the inertia forces.

At a certain Re = Re_s value, laminar regime loses its stability, and the flow becomes turbulent. In a turbulent motion, inertial forces considerably exceed viscosity forces. In this connection, some authors suggest a possibility of neglecting viscosity forces in the study of turbulent flows.

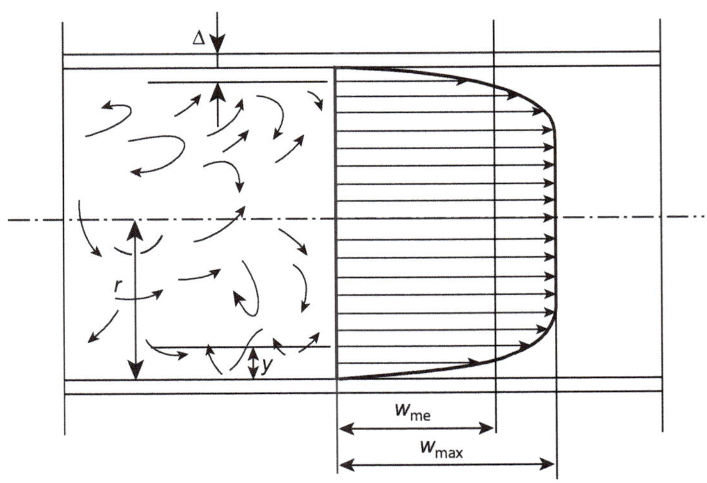

FIGURE 2.6 Flow velocity distribution in a round tube at a turbulent flow regime.

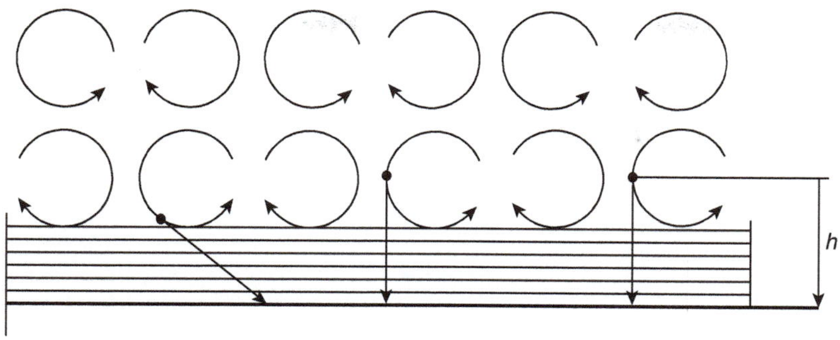

FIGURE 2.7 A pattern of turbulent vortices generation.

Such suggestion simplifies the process, because the system of equations without terms containing viscosity becomes an equation of ideal liquid. However, these equations cannot satisfy boundary conditions of the flow whose velocity must vanish on the walls.

It is assumed that turbulent pulsations arise in the viscous flow zone near the walls confining the flow. Because these zones lag with respect to the core, the mass of medium is periodically emitted into the elevated velocity region (Figure 2.7). Such emissions have a shape of horseshoe-like vortices. The scale of such vortices is comparable with the flow scale, and the velocity and pressure fluctuations frequencies are relatively small. At significant Re number values, the motion of such primary vortices also loses stability, which generates smaller vortices and they, in turn, generate even smaller ones, etc. It is established that the process of vortex formation is a chain process until the smallest vortices arise. The medium motion inside such formations is purely laminar, determined by the molecular viscosity of the medium at a certain temperature. It is noteworthy that the orientation of the main flow does not affect the orientation of small-scale vortices.

Besides the Reynolds number value, turbulent flow regimes are characterized by the following parameters: mean velocity value, turbulence intensity and scale, and fluctuations frequency. A stabilized mean velocity profile is established only at a rather considerable distance from the channel inlet, and not immediately. The medium entering the channel creates a stagnated layer near the walls, which gets thicker as the distance from the outlet increases. A layer with laminar flow is formed, whose thickness is described by the relation:

$$\Delta = S\sqrt{\frac{\nu}{w_{\max}}l}, \tag{2.44}$$

where S is the height of the wall surface irregularities, ν = kinematic viscosity coefficient, l = distance from the channel inlet, and $w_{\max}$ = velocity on the flow axis.

Because of the flow discontinuity, braking against the walls leads to velocity growth in the core, which predetermines a certain elongation of the profile. After achieving a certain thickness, the laminar near-wall layer loses its stability and becomes turbulent. Meanwhile, a thin near-wall layer (viscous sublayer) remains laminar. Its thickness is insignificant and equals, by various sources, from 0.1% to 1.8% of the channel diameter. The viscous layer thickness decreases with growing Re number.

If the boundary layer covers all the irregularities of the wall restricting the flow ($\Delta < S$), then its main core will slide over this layer (Figure 2.8(a)). In this case, the resistance coefficient is independent

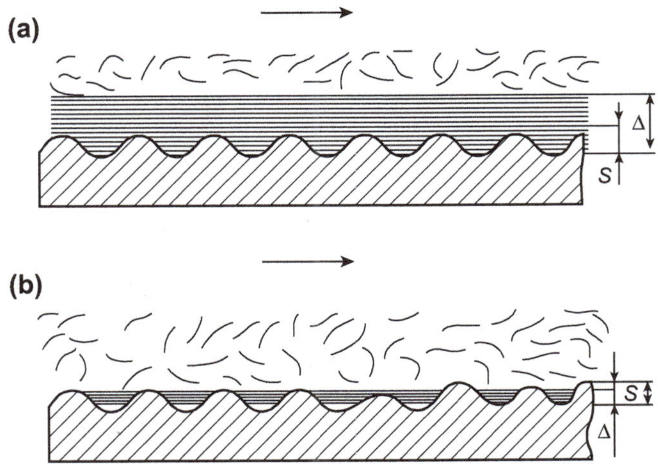

FIGURE 2.8 Different character of flow interaction with a rough wall.

of the walls roughness. With increasing Re, the film gets thinner ($\Delta > S$) (Figure 2.8(b)). The wall surface irregularities get beyond the limits of the boundary layer and interact directly with the turbulent flow increasing losses for overcoming friction. In addition, the phenomena occurring in the boundary layer, in many respects, determine turbulent flows intensity, because they introduce additional disturbing factors into these flows. Hence, it is clear that the resistance coefficient in a turbulent motion depends only on the wall's roughness and is independent of the Reynolds number (Figure 2.9).

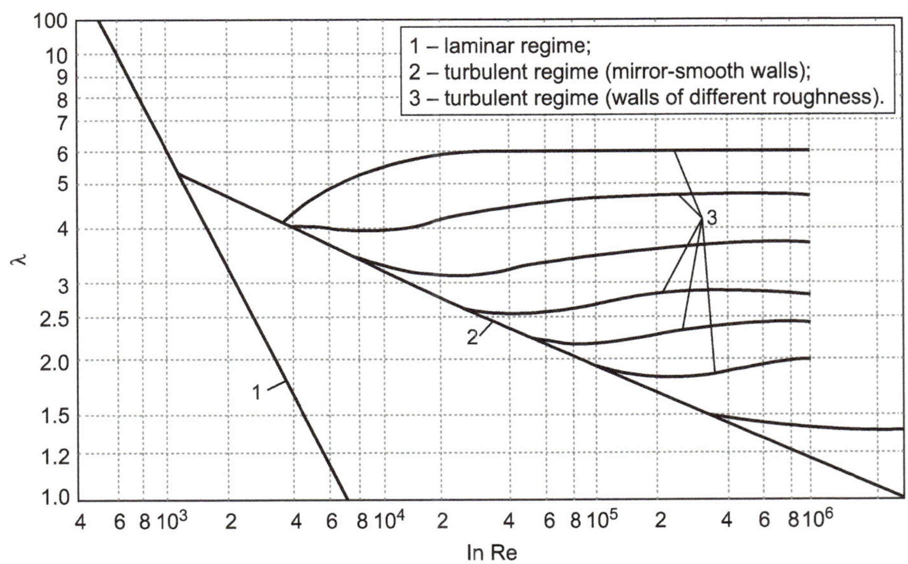

FIGURE 2.9 Dependence of the flow resistance coefficient on the Reynolds number at different wall roughness.

1–laminar regime; 2–turbulent regime (mirror-smooth walls); 3–turbulent regime (walls of different roughness).

For real pipes, a notion of relative roughness is introduced:

$$K = \frac{S}{r}$$

where S is the mean height of irregularities and r is the pipe radius. The resistance coefficient λ remains dimensionless and becomes a function of two variables (i.e., $\lambda = f(\text{Re}, K)$).

The difference between laminar and turbulent flows manifests itself in many phenomena of great importance for numerous technical problems. In turbulent regimes, the impact of a flow on the walls or bodies placed therein is much greater; besides, the medium mixability and flow heat conductivity increase. Thus, the importance of determining the conditions of mutual transitions of both types of flows is perfectly clear.

The value of critical Reynolds number for various specific cases is determined experimentally. For the simplest case of motion along a straight circular pipe, it is established reliably enough that:

$$\text{Re}_s = 2300.$$

However, further studies have shown that Re_s values corresponding to the transition from laminar to turbulent flow in various conditions can differ considerably, being mainly determined by the conditions of the inlet into the apparatus. Experiments are known in which laminar flow regime lasted up to $\text{Re}_s \approx 2000$ and more.

These results show that the Reynolds number in itself is not an unambiguous criterion of turbulence onset. Boundary conditions of turbulence onset should be precisely determined experimentally in each specific case using the relation between the Reynolds number and flow resistance. In a turbulent flow, velocity change observed in the transverse cross-section is smoother than in a laminar flow, except the near-wall layer (Figure 2.6). Because of a complicated character of the process, in this case, the dependence between mean and maximum velocities cannot be derived analytically.

This relation is usually established experimentally in the form of a power expression of the following type:

$$\frac{w_y}{w_{\max}} = \left(\frac{y}{r}\right)^n.$$

Altshul has found that this dependence is more precisely reflected by the relation:

$$\frac{w_y}{w_{\max}} = 1 - 2\frac{\lg\frac{r}{y}}{\frac{0.975}{\sqrt{\lambda}} + 1.35}. \tag{2.45}$$

He determines the ratio between mean and maximum velocities, which is called pipe quality, from the dependence:

$$\frac{w_m}{w_{\max}} = 1 + 1.35\sqrt{\lambda}, \tag{2.46}$$

and the coefficient taking into account velocity nonuniformity over the cross-section – using the expression:

$$\alpha = 1 + 2.65\sqrt{\lambda}. \tag{2.47}$$

Thus, all these characteristics of turbulent flows are determined by the resistance coefficient only and do not depend on Re_s number. Just as for laminar flows, we can determine a layer in a turbulent flow whose velocity corresponds to the mean flow velocity. The distance between this layer and the pipe wall is:

$$y_m = 0.232r.$$

It is rather difficult to compute resistance coefficients for a turbulent flow. Here, we should distinguish three regions:

1. Hydraulically smooth pipes.
   In this case, the near-wall laminar layer exceeds absolute roughness of the pipe ($\Delta > S$). Laminar layer thickness is expressed by:

$$\Delta = \frac{30D}{\mathrm{Re}\sqrt{\lambda}} \tag{2.48}$$

   where D is the pipe diameter. The boundary of the hydraulically smooth pipe region is determined from the relation:

$$\mathrm{Re} \leq 27\left(\frac{D}{S}\right)^{1.14},$$

   and the friction coefficient is determined from:

$$\lambda = \frac{1}{(1.8\lg\mathrm{Re} - 1.52)^2}$$

   and using a more well-known formula:

$$\lambda = \frac{0.316}{\mathrm{Re}^{0.25}}.$$

2. Mixed turbulent regime.
   For this regime, the friction coefficient is determined using an interpolation formula of Colebrook–White:

$$\frac{1}{\sqrt{\lambda}} = 2\lg\left(\frac{K_e}{3.7D} + \frac{2.51}{\mathrm{Re}\sqrt{\lambda}}\right),$$

   where K_e is equivalent or hydraulic roughness.
3. Self-similar quadratic turbulent regime ($\mathrm{Re} > 10^5$; $\Delta < S$).
   Because the thickness of the near-wall laminar layer in a general case is unknown, and, hence, the kind of turbulent regime is also unknown, it is preferable to determine resistance coefficients using a generalized Altshul's formula, which is valid for the entire turbulent flow region:

$$\lambda = 0.11\left(\frac{K_e}{D} + \frac{68}{R_e}\right)^{0.25}.$$

For rectangular channels, mean velocity profile is described by equations analogous to those for circular cross-sections, but with other numerical coefficients. In addition, velocity profiles on wide and narrow walls can differ.

Turbulence scale can be characterized by mean volumes of the medium that take part in the pulsating motion maintaining, for some time, their integrity.

In the near-wall flow region, mixing path length is assumed to be proportional to the distance to the wall restricting the flow.

Turbulence intensity is determined by a value proportional to the ratio of mean square velocity of pulsations to the mean flow velocity.

Pulsations frequency characterizes the change in amplitude values of pulsation velocity. Its value is mainly determined by the vortex scale. Because turbulent flows comprise vortices of various scales, there exists a spectrum of turbulent pulsations frequencies, and not a single frequency.

Pulsational velocity component distribution somewhat differs from the mean velocity. Its maximum irregularity is observed near the walls. The gradient of this velocity in the flow core is not significant.

Here, the relative amplitude of transverse pulsations is not determined by the Reynolds number value even near the walls. At an equal Re criterion, geometrically similar flows are also mechanically similar (i.e., they have geometrically similar configurations of flow lines and are described by the same functions). This is the so-called Reynolds' similarity law. This law is valid for steady motions only, which are not essentially affected by external forces. However, in case of motions essentially depending on external forces or nonsteady motions, the similarity law proves to be more complicated. In such cases, to attain similarity, other dimensionless criteria besides the Reynolds numbers must assume equal values.

It is rather difficult to estimate a reliable boundary for the beginning of purely turbulent regime in apparatuses of arbitrary configuration. It must be found experimentally in each specific case. To do it, one must reveal a dependence of the type:

$$Eu = f(\mathrm{Re}_s)$$

for the apparatus under study on the basis of experimental data, as shown for specific separators in Figure 2.3.

In the region of purely turbulent motion, the resistance coefficient value is independent of the Reynolds number. In the relation under study, the desired dependence is parallel to the axis of Reynolds numbers. Thus, we can reliably determine both the first and the second self-similarity regions. All this points to a complicated nature of phenomena occurring in mono-phase flows.

Many outstanding scientists tried to describe turbulent flow regimes. Poincaré, Heisenberg, Reynolds, Taylor, Karman, Loré, Kolmogorov, Landau, and many more made their contributions to physics and mathematics of turbulent motion, but it still keeps its mysteries. As they say, "turbulence is a junkyard of theories." Recently, a new trend in science called "chaos" has been engaged in turbulence. One result obtained by Feigelbaum is notable, but it has not allowed solving this problem completely. Pulsations and complicated oscillations of separate elements of fluid in a wide range of frequencies make turbulent flow obscure as of yet. Meanwhile, similarity criteria derived on the basis of rather primitive models give engineers a useful tool for estimating and designing complicated hydrodynamic systems, which successfully operate in compliance with estimated predictions.

Probably, scientists will manage to formulate some invariants for two-phase flows, as well, which will allow a correct prediction of their main parameters. In this respect, it is very important to define a mono-phase flow profile, because it defines many aspects of solid phase distribution. The most well-founded empirical relation for the velocity distribution over the cross-section is offered by Altshul:

$$w_r = w\frac{(n+1)(n+2)}{2}\left(1-\frac{r}{r_0}\right)^n \tag{2.49}$$

where w_r is the velocity at the distance r from the flow axis, $n =$ an index depending on the regime of medium motion in a rough pipe, and $w =$ mean flow velocity with respect to channel cross-section. Kogan's relation can be also applied:

$$w_r = \frac{n+2}{n}w\left[1-\left(\frac{r}{r_0}\right)^n\right]. \tag{2.50}$$

The latter relation has a universal character covering all regimes of medium motion. For $n = 2$, the velocity dependence on the radius is parabolic and describes a laminar flow velocity profile; $2 < n < 8$ provides a transient flow regime, and $n > 8$– various stages of turbulent flow.

CHAPTER

Modern Notions of Two-Phase Flows

3

CHAPTER OUTLINE

1. Peculiarities of unit particles settling 37
2. Final settling velocities 43
3. Particle interaction with a moving medium 46
4. Features of two-phase flow generation 49
5. Mass suspension of particles in a flow 52
6. Carrying capacity of two-phase flows 56

1. Peculiarities of unit particles settling

The motion of separate particles in two-phase flows is affected by so many various factors that a general analysis of such motion seems, at first sight, practically hopeless. However, it can be performed step by step, passing from the analysis of simple phenomena to ones that are more complicated.

Modern notions of the two-phase flow mechanism are based on phenomena occurring at the settling of solid particles in an unlimited motionless medium.

Usually, the study of such phenomena implies a strict theoretical analysis of the behaviour of particles of regular shape only, and the settling process is considered as one-dimensional. Possible transverse displacements of particles because of forces similar to lifting force and other factors are not taken into account. Rotary motion of particles during their settling is not taken into account, as well. The motion of particles is considered as a strictly rectilinear one, directed along the gravity force.

If at a certain period of time, the particle velocity is v, then the relative velocity of its overflow by the medium:

$$w = -v \tag{3.1}$$

equals the absolute particle velocity.

At first sight, it may seem that the pressure change in various points of a solid spherical body surface is due only to the fact that it covers a part of space occupied by the medium (Figure 3.1(b)). The simplest possible analytical study of this phenomenon consists in the examination of potential (inviscid) overflow of a sphere.

When passing from point 3 to points 1 and 2 located on the midlength section, the velocity of medium displacement by the settling particle increases. According to the Bernoulli equation, this leads to pressure drop in this section. Behind the midlength section, the pressure starts growing again and

Entropic Invariants of Two-Phase Flows. http://dx.doi.org/10.1016/B978-0-12-801458-5.00003-4

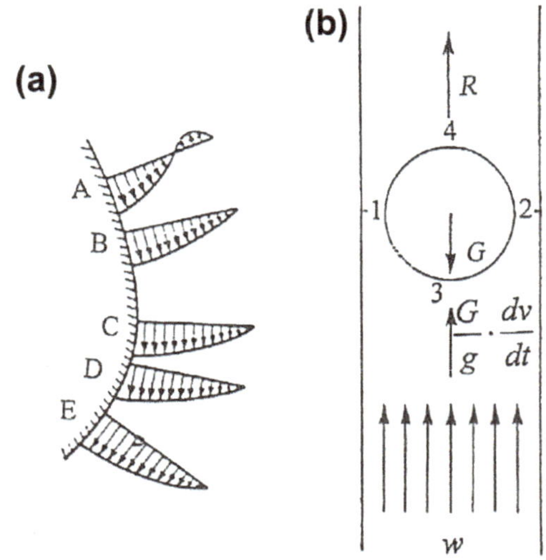

FIGURE 3.1

(a) Particle velocity distribution in the wall layer. (b) Diagram of forces acting on a particle in a flow.

reaches its initial value in point 4, which leads to the formation of a field symmetrical with respect to the midlength section. Hence, the resultant of all forces applied to a particle is zero. Such reasoning leads to a physically impossible conclusion about the absence of resistance at the particle settling. The conclusion based on this reasoning is known as d'Alembert's paradox. This paradox becomes possible owing to excessive schematization of the process under study. Therefore, it is clear, first, that the viscosity of the medium should be taken into account, and, second, it is necessary to examine phenomena occurring on the phase boundary.

Because of the presence of viscosity forces in the liquid or gaseous medium, a so-called boundary layer is formed on the interface. The phenomena occurring in this layer differ greatly from the pattern considered, because energy dissipation takes place in it owing to a considerable change in the medium velocity near the particle surface.

An elementary layer, which moves with the particle, is formed directly on the particle surface owing to cohesion. Because of viscosity force, the velocity is transferred from this layer to nearby elementary masses of the medium. This leads to a monotonic velocity decrease in the boundary layer from v to 0 along the normal to the surface. At that, the velocity distribution curve acquires a characteristic shape reflecting both its continuous decrease starting from the surface and a smooth transition to a motionless medium (Figure 3.1(a)). A negative derivative:

$$\frac{dw}{dy} < 0,$$

which gradually decreases with the distance from the surface and tends to zero while approaching the external boundary of the layer, corresponds to such velocity distribution with respect to the normal. Such situation takes place in the frontal part of a solid or on its entire surface at an overflow without breakaway.

The situation is different in case of motion with boundary layer breakaway. In the region of growing pressure, the medium is dragged not only by the internal friction, but also by the pressure increment along the particle surface, which causes medium motion from higher pressure places to lower-pressure ones (i.e., against the overflow direction). In this case, a certain part of the medium on the boundary layer surface moves in the opposite direction. Therefore, the ratio $\frac{dw}{dy}$ becomes positive within a certain interval on the layer boundary.

Here, the form of the velocity distribution curve is different (A) from that of a flow without breakaway, and becomes regular only as it approaches the surface of the solid. The velocity profile established in the cross-section separating zones of the flow with and without breakaway should represent a limiting curve for the curves of both types (B). In this curve, the inverse flow region collapses to a point for which

$$\frac{dw}{dy_{y=\Delta}} = 0.$$

The location of the boundary layer breakaway is connected with this point. In this place, the boundary layer inflates and breaks away from the particle surface. Hydrodynamic pattern of the process becomes totally asymmetric. As a result, a resultant of the pressure forces arise, which determines the resistance of a settling body in a motionless medium. Although before the breakaway from the surface, the boundary layer was laminar, after the breakaway, it behaves as a free jet in a submerged space and rapidly becomes turbulent. The interface, which represents a surface of tangential discontinuity of the velocity, becomes unstable and rapidly curls into one or several vortices.

Considerable energy dissipation within the whole volume of turbulent wake and interface formation at the separation from the boundary layer lead to the appearance of significant resistance to the particles settling. The resistance is smaller, the narrower the turbulent wake (i.e., the farther on the particle surface is the breakaway point). Thus, viscosity forces are a source of dynamic interaction of two kinds between the solid and the medium. First, these forces manifest themselves as friction resistance at a relative displacement of the particle and the medium. Second, the viscosity of the medium causes the appearance of dynamic forces of backpressure.

The character of particle settling is determined by a system of forces comprising the particle weight in the medium under study and the resistance of this medium (Figure 3.1(a)). The particle weight can be expressed as

$$G = mg_0 \tag{3.2}$$

where m is the particle mass and g_0 = gravity acceleration in a moving medium.

As known,

$$g_0 = g\frac{\rho - \rho_0}{\rho_0} \tag{3.3}$$

where g is gravity acceleration, and ρ, ρ_0 are the densities of the solid and the medium, respectively.

Note that the static lifting force value for air medium is three orders lower than the particle weight, because

$$\frac{\rho}{\rho_0} > 1000.$$

Therefore, without reducing the precision of estimations, we can assume that for a gaseous medium

$$g \approx g_0.$$

For liquid media, on the contrary, this correction should be taken into account, because the specific weights of solid particles and the liquid are of the same order.

In a general form, the resistance of a particle is defined by a relation

$$R = \lambda F \frac{v^2}{2} \rho_0 \tag{3.4}$$

where λ is the particle resistance coefficient, F = midlength section of a particle, and ρ_0 = medium density. The resistance coefficient is an important characteristic determining the total effect of friction forces and dynamic pressure.

Thus, a general equation of a particle motion at strictly vertical settling in a motionless medium can be written as:

$$m\frac{dv}{dt} = -mg_0 + \frac{1}{2}Fv^2\rho_0.$$

A general solution of this equation has the following form:

$$v = -\sqrt{\frac{g_0}{K}}th\left(t\sqrt{g_0K}\right) \tag{3.5}$$

where

$$K = \frac{\lambda F \rho_0}{2m}.$$

Hyperbolic tangent has a limit equal to unity and asymptotically tends to it. Theoretically, this limit is reached at infinity. However, we can assume to a precision sufficient for practice that this function acquires its limiting value at an argument equal to 2.5. Hence, the time of the transition process can be determined as

$$th\sqrt{g_0K} = 2.5.$$

After this period of time, a particle acquires a steady velocity, which is called final settling velocity or falling velocity. It follows from the Eq. (3.5) that

$$v_0 = \sqrt{\frac{g_0}{K}} = \sqrt{\frac{2mg(\rho - \rho_0)}{\lambda F \rho_0^2}}.$$

For a spherical particle, the final settling velocity in an unlimited medium is

$$v_0 = \sqrt{\frac{4gd(\rho - \rho_0)}{3\lambda\rho_0}}. \tag{3.6}$$

It becomes possible to find the final settling velocity value from this relation after determining the resistance coefficient only, because all other values are determined unambiguously.

It is established that all quantitative features of the settling process determining the resistance coefficient (boundary layer thickness as a function of the breakaway angle of the moving medium, breakaway point location, velocity profile in the boundary layer, and the character of its change) depend on the Reynolds number calculated for a particle, for example:

$$\mathrm{Re} = \frac{vd}{\upsilon}$$

where v is particle velocity, d = particle diameter, and υ = kinematic viscosity coefficient of the medium. Experimentally established dependence of the resistance coefficient on the Reynolds number for a sphere is shown in Figure 3.2. A straight line corresponds to the region of rather low Re values (i.e., to the region of laminar overflow without breakaway).

Within this region, self-similarity takes place, and the inverse ratio law is valid (i.e., $\lambda\,\mathrm{Re} = \mathrm{const}$). This region is limited by rather low Re number values. Here, viscosity forces prevail. Stokes determined the resistance coefficient for this region as

$$\lambda = \frac{24}{\mathrm{Re}}. \tag{3.7}$$

This dependence is valid for Reynolds numbers up to 0.2, but it is often used within the range up to 2.

The Stokes law was derived from the assumption that a medium behaves as a continuum (i.e., as a liquid). The process pattern changes at the settling in gas. Here, main relations have a

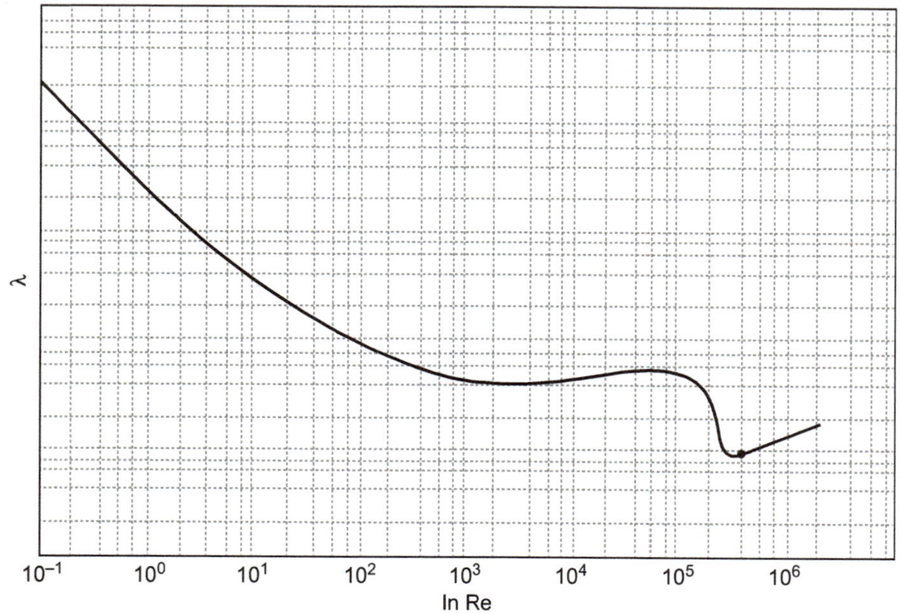

FIGURE 3.2 Dependence of the resistance coefficient of a ball on the Reynolds number at its solitary precipitation in a motionless medium.

lower limit and are applicable only if the Knudsen number for a particle is much smaller than unity, such as.

$$Kn = \frac{l}{d} \ll 1$$

where l is a mean free path of gas molecules. This parameter can be expressed as

$$l = 4.03\frac{\mu}{\rho_0 w}$$

where μ is dynamic viscosity of gas, $\rho_0 =$ gas density, and $w =$ gas velocity.

When the free path of gas molecules is of the same order as the particle diameter, the velocity of the latter becomes greater than according to the Stokes law. This phenomenon takes place for very fine dust or rarefied gas motion conditions. If a particle is very small, its motion does not obey gravitation laws any more, because its collisions with separate gas molecules can displace it to considerable distances in any direction.

As known in practice, the influence of this effect on particles greater than 3–5 μm is negligible. However, the Brownian motion strongly affects the coagulation of such and even coarser particles.

Since the moment of boundary layer breakaway generation, self-similarity is violated. The pattern of particle overflow by the medium becomes much more complicated.

In this regime, the resistance coefficient cannot be determined analytically, because the physical pattern of the overflow is not clear enough. This region is called transitory and covers Reynolds numbers range from 2 to 100, and, according to certain data from 2 to 500. In this region, the resistance coefficient is expressed by various empirical analytical relationships, for example:

$$\lambda = \frac{13}{\sqrt{\mathrm{Re}}}$$

or

$$\lambda = \frac{18.5}{\mathrm{Re}^{0.6}}, \text{ etc.}$$

A deficiency of such formulae is their boundedness by a range of Reynolds numbers and insufficient precision. It is believed that compound formulae are more exact, for example:

$$\lambda = \frac{24}{\sqrt{\mathrm{Re}}} + \frac{4}{\sqrt[3]{\mathrm{Re}}}$$

or

$$\lambda = \pi\left(0.128 + \frac{12.8}{\mathrm{Re}}\right).$$

For $\mathrm{Re} > 1000$, the overflow regime ceases rearranging. A definite form of particle interaction with the medium is established, whose stability is determined by the constancy of the boundary layer breakaway angle ($\alpha \approx 82°$). The value of this angle is stable with respect to changes in Re number, which predetermines the constancy of the resistance coefficient and the presence of self-similarity. The resistance coefficient in this region corresponding to $1000 < \mathrm{Re} < 1 \cdot 10^5$ becomes stabilized and, according to various sources, acquires the values from 0.42 to 0.50.

At such settling, the boundary layer remains laminar. The character of the change in the resistance coefficient in the case of particles settling is much more complicated than in case of the medium motion in a pipe. In the conditions of internal problem, the region of drag coefficient invariability is unlimited from above and extends arbitrarily far over the scale of Re values. In Figure 3.2, the interval of the second self-similar region gives place to a range of a rather strong decrease in λ coefficient. The explanation of this unexpected effect can lie in the character of interaction between the boundary layer and external medium on their interface.

In the second self-similar region, boundary layer turbulization occurs after its breakaway from the particle surface. With increasing Re number, the point of turbulization onset approaches more and more to the breakaway point. At a certain value, turbulization onset overlaps the boundary layer breakaway point (i.e., a part of the boundary layer becomes turbulent). This part is the greater, the higher the Re values. This leads to a rearrangement of the overflow pattern: the separation angle grows up to $\alpha \approx 120°$, turbulent wake behind a particle is significantly narrowed, and the drag abruptly falls. This phenomenon is called drag crisis.

Drag crisis is attributed to an increase in the intensity of momentum exchange between the boundary layer and the rest of the medium. Its onset is the earlier, the higher the overflow perturbation degree (i.e., the smaller the critical value of Re number for the transition to a turbulent regime in the boundary layer).

As follows from this figure, the transition from laminar resistance (Stokes' law) to turbulent one (Newton's law) is gradual in a rather significant range of Reynolds numbers, and not as in a hollow pipe. This is attributed to a relatively large thickness of laminar layer in comparison with particle sizes near their surface. In pipelines, the thickness of near-wall region is negligibly small in comparison to the flow diameter, and, therefore, the transition from laminar to turbulent resistance is more abrupt.

2. Final settling velocities

A particle falling under the action of gravity force in a viscous medium finally starts moving at a constant velocity, with gravity forces equilibrated by hydrodynamic forces. The value of this final settling velocity is accepted, until today, as a determining parameter of the organisation of various processes in disperse materials – classification, pneumatic transport, boiling bed, etc. A great number of theoretical and experimental works deal with finding the value of this parameter.

Equation (3.6) determines final settling velocity of an isolated spherical particle in an unlimited medium.

According to Eq. (3.6), for the Stokes' law validity region,

$$v_0 = \sqrt{\frac{4gd(\rho - \rho_0)\mathrm{Re}}{72\rho_0}}$$

whence

$$v_0 = \frac{d^2(\rho - \rho_0)}{18\mu g}.$$

For the Newton's law validity region, $\lambda = \text{const}$, and the final velocity is determined from the dependence Eq. (3.6). Settling velocity in the transient region is determined by the methods of successive approximations (trial estimations), because the desired velocity v_0 appears as an argument both in Reynolds number and in drag coefficient.

For the transient region, drag coefficient is sometimes determined from an approximating relation

$$\lambda = \frac{A}{\text{Re}^n}.$$

Sometimes, this solution is offered in a criterial form by a relation of the form

$$\text{Re} = f(Ar).$$

For a steady-state motion,

$$\frac{\pi d^3}{6}(\rho - \rho_0)g = \lambda \frac{\pi d^2}{4} \cdot \frac{v^2}{2}\rho_0$$

After respective transformations, we obtain

$$\frac{gd^2}{\upsilon^2 \rho_0} \cdot \frac{d(\rho - \rho_0)}{3} = \frac{\lambda}{4} v^2 \frac{d^2}{\upsilon^2}. \tag{3.8}$$

A dimensionless complex

$$\frac{gd^3(\rho - \rho_0)}{\rho_0 \upsilon^2} = Ar \tag{3.9}$$

is called Archimedes criterion. Taking this into account, we can transform Eq. (3.8) into

$$\frac{Ar}{3} = \frac{\lambda \text{Re}^2}{4}. \tag{3.10}$$

Substituting the value of λ corresponding to various overflow regimes into Eq. (3.10), we obtain:for laminar regime:

$$\text{Re} = \frac{1}{18} Ar;$$

for turbulent regime:

$$\text{Re} = 1.742 Ar^{0.5};$$

for transient regime:

$$\text{Re} = 0.153 Ar^{0.174}.$$

A universal formula connecting these two criteria for all overflow regimes is suggested by:

$$\text{Re} = \frac{Ar}{18 + 0.6\sqrt{Ar}}. \tag{3.11}$$

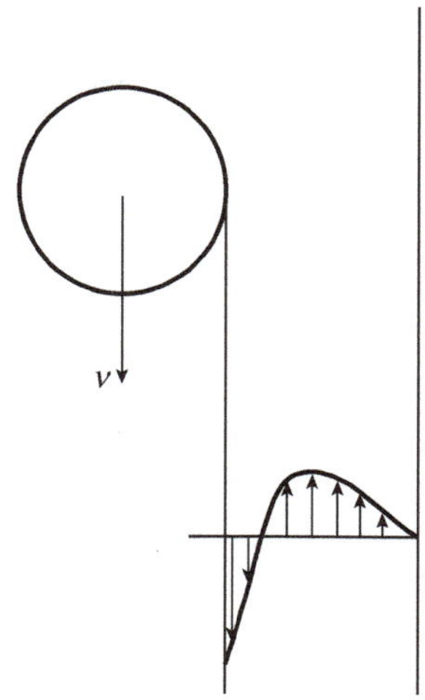

FIGURE 3.3 Influence of a wall on a particle precipitating in its vicinity.

Settling of isolated particles in a medium limited by solid walls has some peculiar features. A solid particle overlaps a part of the space, and therefore the medium it displaces moves in the inverse direction. The character of velocity distribution profile near the wall is shown in Figure 3.3. The general character of particle interaction with the wall depends on the particle shape and size, position and orientation, as well as on the wall geometry.

Thus, a rotational motion of the medium arises around a particle settling in a limited space. Volumes directly adjacent to the particle move along with the latter, while those located at a distance beyond the boundary layer move in the opposite direction. This leads to increasing drag and decreasing settling velocity.

At a turbulent motion, wall impact on the particles settling is somewhat decreased, but, nevertheless, the effect is not negligible, because it considerably corrects the actual velocity of each separate particle settling in comparison with estimations, especially at mass settling.

In certain conditions, the restraint determines phenomena connected with both particles migration in the direction perpendicular to settling and their rotation.

Particles of spherical and irregular shapes behave differently. In the laminar overflow regime, particles of irregular shape fall, in most cases, in the position they are introduced into the medium. Here, their velocity depends on the initial orientation, and, therefore, identical particles can have different settling velocities. In the transient region, particles orientation becomes unstable and is accompanied by oscillations, whose amplitude increases with growing Re number. In the turbulent region, particles fall irrespective of their initial orientation and hover in a position ensuring maximal drag coefficients. Particles of irregular shape have a more marked tendency to rotation and transverse migration at the settling. Thus, the process of simple settling of even separate particles in a motionless medium is very complicated and uncertain. Naturally, the behaviour of a particle in moving media is characterised by much more complicated and uncertain phenomena.

3. Particle interaction with a moving medium

In comparison with settling, in this case, external turbulence leads to a significant decrease in drag coefficient of a particle because of a decrease in the upper critical value of the Reynolds number at a simultaneous displacement of the annular line of the flow breakaway from the sphere to the stern. Such displacement is already observed for $Re>400$. At lower Re values, this effect is weakened. Breakaway point is not displaced, but a certain increase in the drag is observed at the expense of increasing energy dissipation in the wake region.

If a particle is small in comparison with the smallest turbulence scale, it responds to all pulsations inherent to turbulent motion. This feature can serve as a basis for the primary definition of the difference between the behaviour of coarse and fine particles in a flow. Coarse particles participate, for the most part, in a linear motion of the medium, whereas fine particles follow turbulent vortices. The resistance to fine particles motion is determined by the viscous nature of the medium. Although the velocities of particles and the medium differ by the slip value, the presence of particles in a flow enhances dissipation. This predetermines exceptional complexity of particles motion, which hardly keeps within the frames of cellular model suggested by Chen. According to the latter model, a particle moves together with a certain volume of deformed medium. This approach can be satisfactorily applied, with many reservations, to laminar flow regimes only.

It is difficult to understand the mechanism of the phenomenon under study because of the appearance of secondary motions, particles oscillations, and rotations, which considerably affect drag coefficient. As noted by Sow, oscillations are not observed at $Re<80$ and always take place at $Re>300$. Particles rotation in a liquid flow can be caused by various reasons.

If a particle is in a gas having velocity gradient, it is set in rotation. Although shear velocity in turbulent vortices can be high, this effect is self-compensated because of the random nature of turbulence and exerts insignificant effect on particles rotation. The flow near the walls is an exception. In this layer, at the fluid or gas motion, the mass of the medium joins a rotating particle, which leads to the flow velocity increase on one side of the particle and its decrease on the other. This phenomenon known as Magnus effect makes the particle shift into the region with a higher velocity (toward the flow axis). However, according to precise experimental data, particles are concentrated in an annular layer at a distance of about a half of the pipe radius from its axis. Transverse impact of the flow can also arise at the expense of the displacement of the boundary layer breakaway point at the particle rotation.

It should be admitted that this phenomenon is, in general, extremely complicated, and only idealised cases have been studied. The results of such studies are useful, because they reveal comparative importance of various factors. Busroid has shown that, in a laminar flow (or laminar boundary layer), the ratio between the transverse force and the force determined from the Stokes law equals

$$\frac{F_n}{F_c} = \frac{0.121d}{D} Re_n f$$

where f is friction coefficient at the gas motion.

The analysis of this ratio shows that the tendency of particles motion to the direction normal to the wall is marked enough. Such phenomena are often used in practical purposes when the layer of the fluid flow is moderate, for example, at the concentration in plates, tables, etc. The force transverse to the flow predetermines in many respects the character and results of separation. It has been noted that, in certain cases, at Re ≈ 10 a particle moves to the flow axis, whereas at 16<Re<120 it moves toward the wall. As we can see, the problem of particle interaction with a moving medium is far from being solved yet.

However, it is necessary for practical purposes to study systems containing a great number of particles. Until recently, two-phase flows have been studied at the level of determining hovering velocities of particles in an ascending flow. This is done without taking into account particles rotation, their transverse migration, absence of collisions between particles and the wall, thus reducing a complicated process to a linear one-dimensional problem. Here, the medium flow velocity is accepted as deterministic and equal over the whole channel cross-section. In such extremely idealised conditions, relative velocity of a particle overflow is determined by the Eq. (3.1). A general equation of a spherical particle motion in these conditions is

$$m\frac{dv}{dt} = -g_0 + \frac{1}{2}\lambda F\rho_0(v-w)^2.$$

This expression can be transformed into

$$\frac{dv}{dt} = -g + K(v-w)^2$$

where

$$K = \frac{\lambda F\rho_0}{2m}.$$

In such form, this dependence represents a Riccati-type equation reducible to a differential equation of the second order. The solution of this equation gives

$$v = w - \sqrt{\frac{g_0}{k}}th\left(t\sqrt{g_0 K}\right). \tag{3.12}$$

The second factor in Eq. (3.12) represents a hyperbolic tangent asymptotically tending to its limit. After a certain time interval, the particle velocity becomes practically constant and later on is independent of time and determined by a relation

$$v = w - \sqrt{\frac{g_0}{K}}.$$

This value is called a steady particle velocity and is determined by the flow velocity only.

A limiting case in which the steady velocity theoretically equals zero is the most interesting. The medium flow velocity ensuring this condition is called hovering velocity and is determined by a relation

$$w_0 - \sqrt{\frac{g}{K_0}} = 0$$

where from

$$w_0 = \sqrt{\frac{4gd(\rho-\rho_0)}{3\lambda\rho_0}}. \tag{3.13}$$

Based on the obtained result and Eq. (3.6), it is usually concluded that the final settling velocity and hovering velocity in the counter-flow for a spherical particle are identical. The dependence of Eq. (3.13) is a consequence of an equation obtained by excessive idealisation of the phenomenon. At that, it is implicitly assumed that drag coefficients of settling and hovering particles are the same and independent of the Reynolds number (i.e., of the turbulence of the medium), which is present in a flow and absent in a quiescent medium.

The relation between hovering and settling velocities of the same particles is a very important problem. Reliable data on the possibility of determining one characteristic parameter through another for all ranges of drag change are practically absent. Therefore, hovering velocity values for each specific case are determined experimentally, mainly, within the segments of stabilized motion. Various experimental methods and approaches are used for this purpose – visual, photoelectronic, tagged particles, rapid filming, instantaneous intercepting of channel segments, etc.

It is assumed that drag coefficient of particles increases with growing acceleration, and the effect of the latter on this coefficient value can be rather significant. It is established that for gaseous media with $\frac{\rho}{\rho_0} > 1000$, drag coefficient is independent of the acceleration sign and equals the value of this parameter for a sphere overflown at a constant velocity.

The above pattern of the particle and flow interaction becomes much more complicated at a transition to particles of irregular shape. The behaviour of such particles in a flow is, as a rule, unsteady, with a marked tendency to rotary motion and migration. In most commercial powders, the particles shape is such that they do not have a symmetry axis. Therefore, the impact of a flow generates a resultant moment whose value is unstable. This causes the appearance of elevated Magnus forces.

The behaviour of particles of irregular shape is practically completely unexplored. The issue of the form-factor of such particles remains debatable and far from being unambiguous until now.

Urban classifies all particles of irregular shape into two groups. The first group comprises particles for which the boundary layer breakaway is unambiguously stipulated by the presence of overflow edge. The shape of such bodies or particles is characterised by either increasing or constant cross-section before the overflow edge and its drastic change on the latter. Drag coefficients of such bodies are independent of Re number over the entire range of its variation. They represent rectangular plates, small cylinders, cones, and hemispheres oriented in a corresponding manner with respect to the flow.

The second group comprises rounded, poorly streamlined bodies without sharp edges, whose cross-section decreases gradually in the flow direction. Here, the place of boundary layer breakaway is stipulated by the overflow character, and, therefore, drag resistance depends on the Reynolds number.

Drag coefficients of particles of irregular shape are reduced by various methods to a respective characteristic of an equivalent sphere. The form-factor is assumed as a ratio of drag coefficients of the body and the equivalent sphere.

$$K_g = \left(\frac{\lambda}{\lambda_0}\right)_{d_0=idm;\mathrm{Re}=idm}.$$

The more the particle shape differs from a sphere and the coarser its surface, the higher its drag coefficient and the lower its hovering velocity. It is established that drag coefficients of particles of irregular shape depend not only of their geometry, but also on the Reynolds number, for example:

$$\lambda = K_g\lambda_0 = f(K_g;\mathrm{Re}).$$

Obviously, there exists a relation of the following sort between dynamic and geometrical form-factors:

$$K_g = f(K; \mathrm{Re}).$$

For particles of irregular shape, this dependence has not been sufficiently studied, and the problem of determining drag coefficient of such kind of particles remains rather disputable. What is known is that in the transient region, the dynamic form-factor increases with growing Re. This points to a stronger dependence of drag coefficient on Re for irregular particles than for a sphere.

Self-similarity region for particles of irregular shape ($\lambda = \mathrm{const}$) comes at lower Reynolds numbers. This shift is the greater, the higher the geometric form-factor. This points to the fact that turbulization starts earlier at the surface of non-spherical particles than at the surface of a sphere.

As Uspensky has shown, particles of irregular shape are oriented in a flow in such a way that their resistance becomes maximally possible. This promotes early turbulization of the medium in the stern region. Gorbis has confirmed this fact by his experiments with aluminium cylinders.

Particle roughness also affects drag coefficient. Particles with a rougher surface, other conditions being the same, have a lower hovering velocity. The effect of particle roughness, especially if they are of irregular shape, is insufficiently studied, and it is practically impossible to consider this factor in theoretical estimations, especially in case of fine particles.

If w_0 or λ are determined experimentally, the use of these parameters in analytical estimations provides the account for all side effects. Therefore, we can write

$$mg_0 = \lambda F \frac{w_0^2}{2} \rho_0$$

for experimentally determined hovering velocity, that is, particle resistance equals its weight in the medium and, at the same time, particle velocity $v = 0$, for example:

$$\frac{\lambda F}{2} \rho_0 = \frac{mg_0}{w_b^2}.$$

In a general case, particle drag

$$F = \lambda F \frac{(w - v)^2}{2} \rho_0.$$

Considering this, an interesting relation is obtained containing all particle characteristics in the form of parameters that can be easily determined experimentally, which does not contain drag coefficient in an explicit form:

$$R = \frac{mg_0(w - v)}{w_b^2}.$$

4. Features of two-phase flow generation

Joint settling of a group of particles is accompanied by a displacement of a large amount of medium, which causes intense turbulent motion of liquid drawing in a lot of particles, even those moving upward (Figure 3.4).

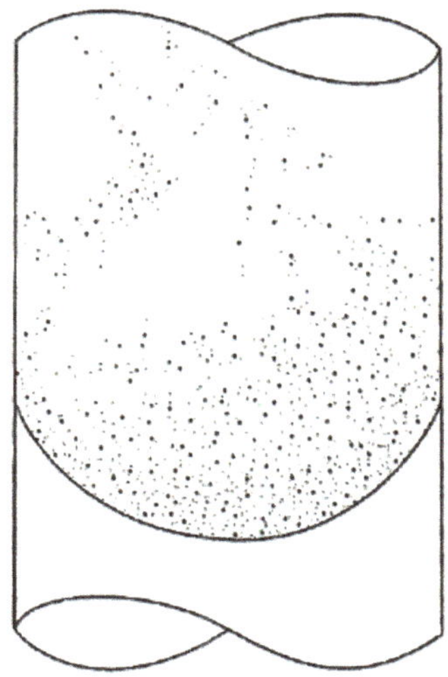

FIGURE 3.4 Nature of mass precipitation of solid particles in water.

At such settling, significant local concentrations in the core and non-uniform concentration in the remaining parts of the cross-section are formed. Finally, the particle core gets destroyed by erosion owing to the medium resistance, and its falling velocity decreases in compliance with the decrease in its size.

The core becomes cup-shaped, because the particles located above the phase boundary in it move faster than those located in the lower part. Such a core can overtake a smaller cloud of particles and absorb it, increasing in size as a result.

This phenomenon puts very serious questions regarding the organisation of material feed into the apparatus for separation. Apparently, it should be fed in a maximally aerated form and periodically, with the cycle ensuring one portion distribution before feeding another. If the material is fed into the apparatus in a compact state, one needs mechanical devices for destroying the core of particles and idle apparatus height for the erosion of its remaining parts by the medium flow.

This phenomenon could be avoided at extremely low concentrations of fine particles. In the region of core destruction, solid particles concentration in the cross-section is levelled. Here, the resistance force acting on a particle is more significant than in case of a solitary particle because of the following two reasons:

1. Velocity gradient near the particles increases because of the proximity of other particles.
2. In the case under study, upward displacement of the medium arises, which further increases the velocity gradient.

At a sufficient particles concentration, the interaction of boundary layers of neighbouring particles starts playing an important role. Naturally, this phenomenon leads to drag coefficient change. According to (3.13), we can write for the flow core:

$$v_a = \sqrt{\frac{4gD(\rho_c - \rho_0)}{3\lambda_a \rho_0}}$$

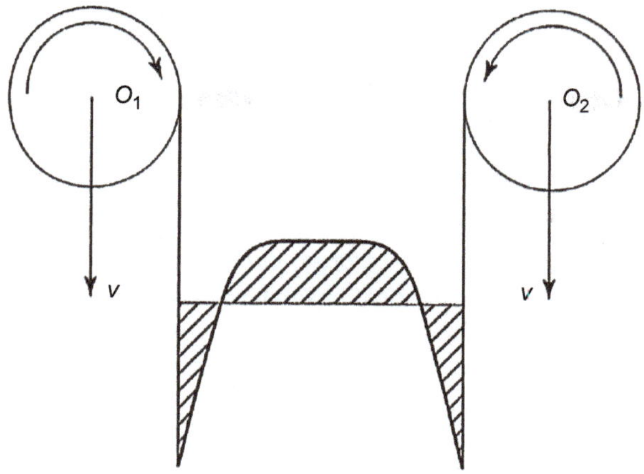

FIGURE 3.5 Schematic diagram of coprecipitation of two identical particles.

where D is the pipe diameter, ρ_c = bulk density of particles in the core, and λ_a = drag coefficient at the suspension settling.

For a particle inside the core,

$$v = \sqrt{\frac{4gd(\rho - \rho_0)}{3\lambda\rho_0}}.$$

For $v_a > v$ a core is formed, while for $v_a \approx v$ particles move independently. Happel and Brenner describe experiments modelling the settling of particles with uniform concentration. Two identical particles settling in parallel rotate toward each other. In this case, settling character is determined by the conditions shown in Figure 3.5. The shaded area shows the distribution of vertical medium velocity along the centreline O_1O_2. If two identical particles settle vertically one after another, the back particle acquires a higher settling velocity and overtakes the front one. The resulting doublet increases its velocity.

Many studies of compact settling in uniformly dispersed systems have been carried out. A number of empirical relations comprising the volume fraction occupied by particles have been suggested, the most well-known among them being the following.

Hirst and Lyaschenko expressed the correction factor taking into account the effect of concentration on the settling rate by an empirical relationship:

$$\varphi = (1 - \beta)^n$$

where β is bulk concentration of the solid phase, and n = experimental factor.

Mints and Schubert have established, on the basis of experiments with gravel and sand that the value of this parameter varies within the limits of 2.25–4.6.

They have clearly shown that this factor cannot be constant and depends on the particles overflow regime.

Kachan defines hovering velocity in compact conditions by the parameter:

$$\varphi = \left(\frac{1 - \beta}{\beta}\right).$$

Todes has suggested, on the basis of experimental data generalisation, a common relationship for all particles overflow regimes:

$$\mathrm{Re} = \frac{Ar(1-\beta)^{4.75}}{10 + 0.6\sqrt{Ar(1-\beta)^{4.75}}}.$$

The value of concentration-dependent factor is greatly affected by the shape of particles, this impact manifesting itself differently under free and compact conditions.

We can conclude from the available publications that for low concentrations $\mu \leq (1 \div 1.5)$ kg/m^3, the value of the parameter φ can be assumed equal to unity to within 5%. However, experiments have shown that the influence of concentration manifests itself at such values in the appearance of a core of particles settling together, and not in the interaction (either direct or through boundary layers) of uniformly dispersed particles with each other. This phenomenon is also observed and should be taken into account at much lower consumed concentrations.

5. Mass suspension of particles in a flow

Until recently, in an overwhelming majority of cases, the mechanics of two-phase flows has been dealing with monodisperse composition of the solid component.

However, regularities revealed as a result of such study cannot be used for obtaining qualitative and quantitative relations applicable to a real process with the solid phase representing usually a polydisperse material with a rather broad size range. Such composition of the solid phase essentially alters hydrodynamic conditions of the process because of the appearance of new phenomena, which are absent in flows with mono-dispersed material. The number of works studying two-phase flows with a polydisperse composition of the solid component is rather limited, although, recently, this problem has been attracting serious attention.

At the analysis of regularities of ascending flows with a polydisperse substance, new important aspects of the flow have been revealed. The main one is related to particle collisions in a flow and formation of agglomerates moving as a whole. These two mechanisms are interconnected and, to a greater extent, are a consequence of polydisperse character of particles and turbulence of the two-phase flow. The formation of aggregates is the most marked in suspensions in gases. This phenomenon arises because of a permanent drawing of the surrounding medium into the turbulent wake formed behind a moving particle. Particles move faster toward the hydrodynamic wake because of the origination of a local pressure drop. This leads to the formation of a conglomerate of two or more particles. The formation of conglomerates is also promoted by local pressure non-uniformities, which are the most pronounced for a turbulent flow. At a mutual approach of two or more particles in a flow, the velocity of their interdirected motion grows at the expense of instantaneous narrowing of the distance between them.

Interaction of solid polydisperse particles in a flow is an extremely complicated physical process. Colliding particles can either agglomerate or diverge having exchanged their momentums. If before the collision, the particles represented aggregates, then after the collision these aggregates can be either totally or partially destroyed or, on the contrary, grow. A collision occurs mainly owing to the different velocities of the solid particles motion.

Relative velocity of such particles (and their clusters) may be due to various reasons—size, aggregates configuration, character of local turbulent vortices, etc.

This can be attributed to the complicated character of the phenomenon, which is obviously impossible to study experimentally, because any contact device in a flow affects the state and behaviour of aggregates of particles. Sometimes, the influence of agglomeration is insignificant, for example, in gaseous suspensions with coarse and granulated particles. In flows with the particle size below 60 μm, this phenomenon is, on the contrary, extremely marked; it is stronger the smaller the particles.

Sometimes aggregation is used with a positive effect, for example, for dust collection. It is known that coarse fast-falling particles can remove finer ones out of the suspension. Fux named this phenomenon "kinematic coagulation."

Mechanical interaction of particles in a flow was confirmed by a very simple experiment. In the apparatus whose schematic diagram is shown in Figure 3.6, air flow velocity was chosen in such a way that a suspended layer of spheres 12–15 mm in diameter and with $\rho = 6000\ \mathrm{kg/m^3}$ was formed in the conical part of the apparatus. The layer was stabilized over the whole cone height. After that, fine coal dust ($d < 0.25$ mm) was fed into the air flow in such small amounts that the air transparency did not deteriorate ($\mu < 0.1\ \mathrm{kg/m^3}$). Under the action of this dust, massive heavy balls suspended inside the cone were thrown out into the cylindrical part of the apparatus to the height 400 mm from the cone edge. This experiment visually demonstrates the character of the effect of particles interaction.

For identical regime parameters, the frequency of particles interaction essentially depends on physical properties of the material and, in the first place, on the particles elasticity. The number of interactions between particles of different sizes increases with growing material concentration in the flow. Obviously, the character of the motion of any fraction of the material to be separated in a flow is closely connected with size distribution of particles of other classes.

At the collisions of particles and aggregates, they exchange momentums. At that, coarse particles are accelerated in the flow direction, and fine particles are slowed down. A conclusion follows that at significant concentrations, all particles of a polyfractional mixture somewhat level off their velocity,

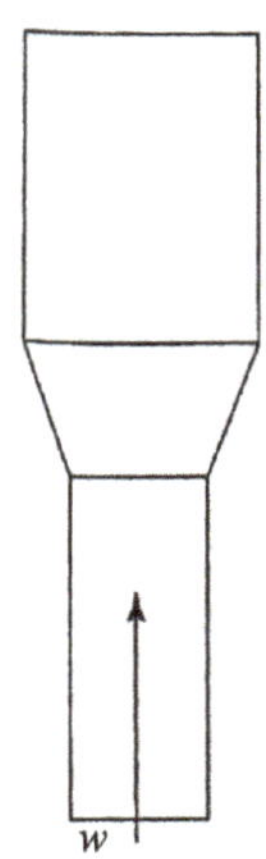

FIGURE 3.6 A facility for suspending coarse heavy particles in the counter-current.

which is observed, for example, in vertical pneumotransport. However, there is an essential difference in the behaviour of suspensions of coarse and fine particles.

It was assumed in the studies of the resistance to two-phase flows in pipes that the introduction of a solid component leads to an increase in pressure losses. For a moderate ratio between the flow rates of the solid and gaseous phases, Gastershtadt has introduced a relationship:

$$\Delta P = \Delta P_0(1 + K\mu)$$

where ΔP is the two-phase flow resistance, $\Delta P_0 =$ the pure air resistance, $\mu =$ solid phase concentration, and $K =$ coefficient.

For many years, further researchers believed that this statement was unshakeable. Urban canonised this statement by introducing additivity principle for pressure losses in a pure flow and solid phase separately. However, we have experimentally established that, under certain conditions, pressure losses in a pipe flow decrease at the addition of solid particles to a lower level than in a pure flow. Such a phenomenon is characteristic of fine particles only, it is never observed in flows with coarse particles. It is established that, at solid particles concentration causing said effect, the velocity profile of a gaseous medium remains practically unchanged because of the presence of solid particles. In this case, the average concentration of particles is such that the distance between them exceeds their diameter 10 times and more. An unambiguous explanation of this effect is not found yet.

A dual character of particles impact at both generation and suppression of turbulence has been established. Ribender and Reyner point out that if particles cannot follow the motion of vortices, they stabilize the flow and create conditions for a laminar flow.

Particles are unable to follow medium slow-down in the boundary layer. For apparatuses extended throughout the height, this leads to a significant increase in particles concentration near the wall, and their transfer to this layer is mainly realised owing to turbulent diffusion in the flow core. Besides, here the mechanism of particles overshooting low-turbulence region owing to their exit out of the surrounding vortex because of their inertia also takes place. As a result, various velocity profiles for the solid phase can set in an ascending gas flow, the most characteristic ones being shown in Figure 3.7.

All this determines the conditions that, at average ascending flow velocities sufficient for fine particles removal, a part of them go out downward, against the flow, while a part of coarse particles, whose hovering velocity sometimes considerably exceeds the average velocity of the medium, is carried out upward, into the fine product.

An effect contributing to a permanent material output to the channel walls is also observed. It is noted that after getting into the near-wall region, a particle can be longitudinally displaced along the walls without leaving this region. It is established that such a motion is short-term at an ascending motion of a particle and rather long-term at its descending motion.

Because of the counter-directed migrations of particles, whose causes are examined in detail, the concentration maximum is sometimes formed somewhere in the middle between the flow axis and the wall (Figure 3.8), and not directly near the wall.

To simplify the understanding of the phenomena occurring in two-phase flows, they are considered sometimes as a single-phase pseudo-homogeneous medium with elevated viscosity and density. Such an approach is insufficient for an appropriate reflection of the main phenomena in dispersed flows, because it is basically pseudo-physical, reducing two-phase flow mechanism to the single-phase medium flow.

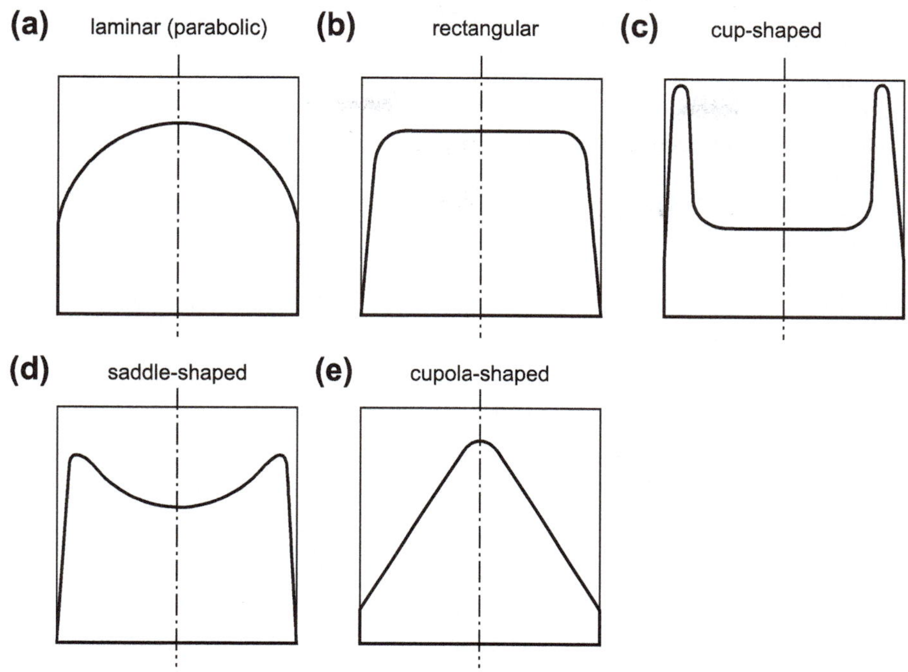

FIGURE 3.7

Various experimentally obtained velocity profiles for the motion of solid phase of different compositions in an ac sending flow: (a) laminar (parabolic); (b) rectangular; (c) cup-shaped; (d) saddle shaped; and (e) cupola-shaped.

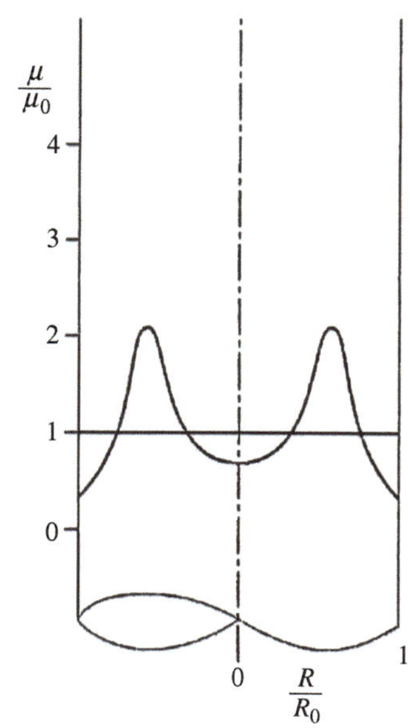

FIGURE 3.8

Cup-shaped distribution of narrow-class particles concentration in a flow: μ_0 = concentration on the flow axis; R_0 = channel radius.

We can agree with Bart's conclusion that the Reynolds number plays a secondary role in two-phase flows. For such flows, besides the final settling velocity and hovering velocity, another parameter is determined—minimal transfer velocity. This velocity represents the minimal flow velocity at which nothing settles against the current. It can be easily shown that this velocity is somewhat higher than the hovering velocity of the respective particles.

It will be realistic to calculate mass transfer processes in flows only if we find conditions for determining regimes providing any partial removal out of the apparatus and, as a particular case, minimal transfer velocity.

Thus, the formulated problems are very important for the study and understanding of the mechanism of two-phase flow. For their solution, extensive experimental and theoretical researches are needed.

6. Carrying capacity of two-phase flows

The objective of the present research is to reveal the force impact of a two-phase flow on a fixed sensor in the apparatus cross-section and the profile of force impact of the continuous medium in the two-phase flow. A pipe of circular cross-section was chosen for the study.

The measuring system is presented in Figure 3.9. Four holes 6 mm in diameter are made in a round tube at the distances 50, 250, 450, and 750 mm from the lower edge. A probe 2 (glass tube Ø = 4.7 mm displaceable along its axis) is installed by turns into these holes to a fixed depth $X = 100$, 95, 85, 75, 65, 55, 45, and 35 mm. Force impact of the continuum perceived by the probe is transferred by a lever facility 3 using a steel needle 4 to scales with the measurement range from 0 to 500 g and scale interval of 0.1 g. These scales register the measuring system reaction to disturbances in a continuous regime. Movable table 6 is used for the probe centring at its displacement with respect to the hole axis.

The experiments were performed on periclase ($\rho_m = 3600$ kg/m^3). Consumed concentration of the material was maintained at the level of $\mu = 1.5$ kg/m^3. The amount of experimental material was 2.5 ÷ 3 kg. Granulometric analysis was performed on a set of sieves with the mesh size 0.75, 0.5, 0.3, 0.2, and 0.14 mm. The hole for material feed into the apparatus was located at the height of 350 mm from the lower edge of the pipe. The total pipe height was 1200 mm. Air was fed into it from below at the velocities of 2.86, 3.96, 4.92, 5.66, and 6.3 m/s.

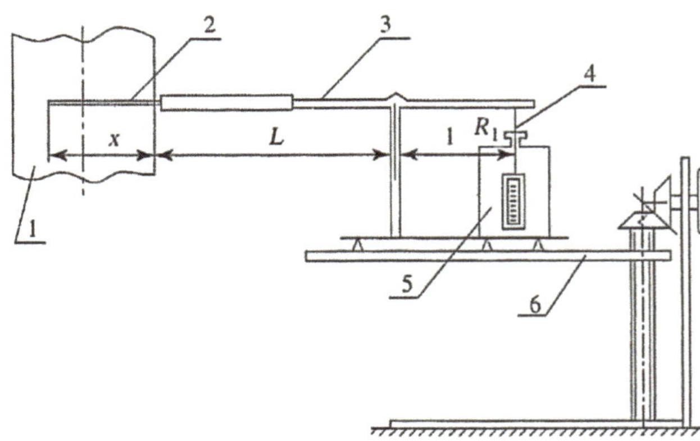

FIGURE 3.9

Experimental facility layout: 1 = channel; 2 = probe; 3 = lever facilities; 4 = needle; 5 = scale; and 6 = mobile table.

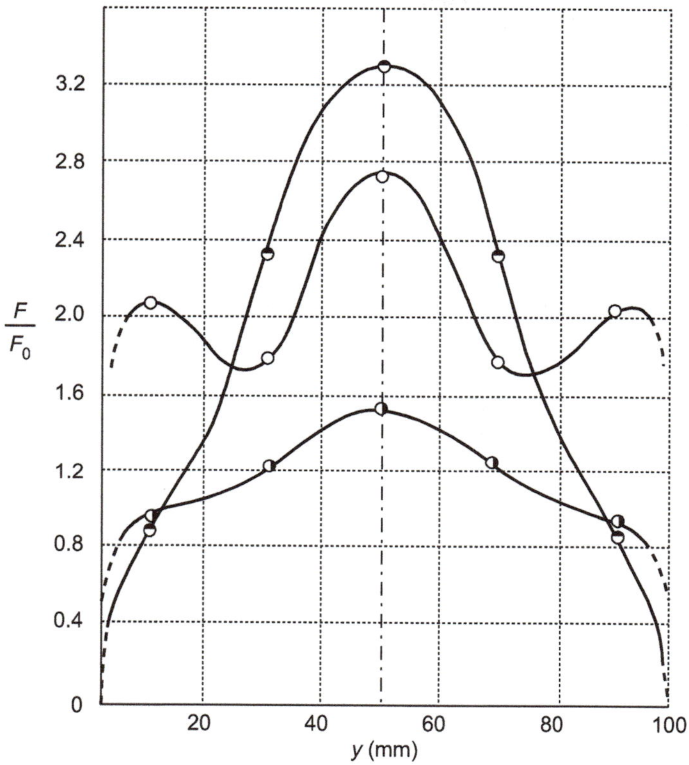

FIGURE 3.10

Carrying capacity profiles of a two-phase flow for $w = 3.96$ m/s; $\mu = 1.5$ kg/m^3. Distance between the measurement point and the lower channel edge: 100; 300; and 700 mm.

The obtained profiles of the carrying capacity of a two-phase flow are of similar character. Figure 3.10 shows the result for flow velocity $w = 3.96$ m/s at the consumed material concentration $\mu = 1.5$ kg/m^3. As follows from Figure 3.11, the carrying capacity of a two-phase flow drastically differs from the carrying capacity of pure air. Thus, for example, it can exceed the effect of pure air on

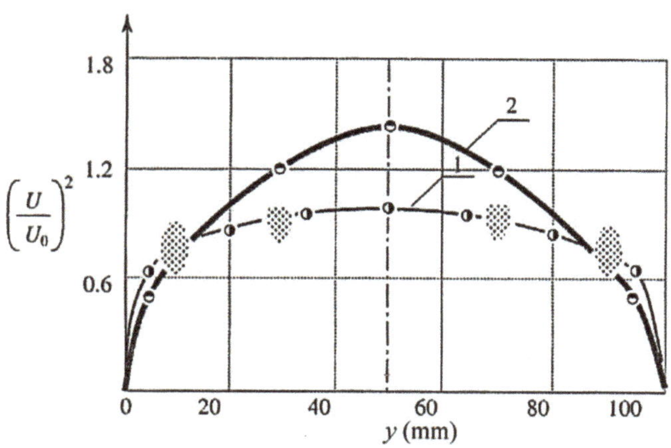

FIGURE 3.11

Carrying capacity profiles: 1 = clean air; and 2 = two-phase flow.

the axis three and more times. Profile of the carrying capacity of a two-phase flow is formed depending on the measurement point. A sharper profile and maximal carrying capacity on the axis are characteristic of the upper parts of the apparatus. The lower the segment position in the apparatus, the smaller is relative carrying capacity on the flow axis. Pay special attention to the lower curve obtained at the point close to the place of material feed into the apparatus. This profile is extremely non-uniform, which is caused by high local material concentrations and its non-uniform distribution over the cross-section of the apparatus.

Carrying capacity profile of a solid phase in a two-phase flow in classification conditions is significantly deformed and becomes more peaked.

The performed analysis of the process mechanism and of the degree of insight into its physics clearly shows that at present there are no prerequisites as yet for the analytical study of the problem. Therefore, its main integral regularities are usually established only experimentally. This approach allows revealing such regularities at a purposive study of the phenomenon. Unfortunately, the available extensive empirical material obtained, as a rule, on various apparatuses, does not contain the necessary generalising regularities.

CHAPTER

4 Empirical Invariants for Two-Phase Flows

CHAPTER OUTLINE

1. Fractional extraction 59
2. Affinization of fractional extraction curves 64
3. Unified separation curves 68
4. Generalizing empirical invariant 74

1. Fractional extraction

Consider bulk material fed into an ascending flow (Figure 4.1). At high velocities, all the material is lifted by the flow, thus realizing a transporting regime. At low velocities, all the material falls against the flow, thus realizing a so-called falling layer. In these two limiting regimes, mass distribution does not occur, because all the fractions move together in the same direction. The situation is different when the flow velocity value lies between these two limiting values. Figure 4.2 shows different fractions distribution in such a flow. Bulk material usually consists of particles with a broad range of sizes. A narrow size class or a fraction implies particles within a size range in which the ratio between the finest ($x_{\min}$) and the coarsest ones ($x_{\max}$) is

$$\frac{x_{\max}}{x_{\min}} \leq 2$$

To simplify the particle size determination, which reduces experimental error, bulk material of elevated fineness is the most suitable. In our example, the fractions ranges were as follows: (10–7) mm, (7–5) mm, (5–3) mm, (3–2) mm, (2–1) mm, (1–0.5) mm, and (0.5–0.25) mm.

A narrow class size is the arithmetical mean of limiting sizes of particles in a fraction, i.e.

$$x = \frac{x_{\max} + x_{\min}}{2}$$

Fractional extraction for every narrow size class, in percent, is determined by dependencies of the following type

$$\left.\begin{aligned} F_f(x) &= \frac{r_f}{r_s}\gamma_f \\ F_c(x) &= \frac{r_c}{r_s}\gamma_c \end{aligned}\right\} \tag{4.1}$$

Entropic Invariants of Two-Phase Flows. http://dx.doi.org/10.1016/B978-0-12-801458-5.00004-6

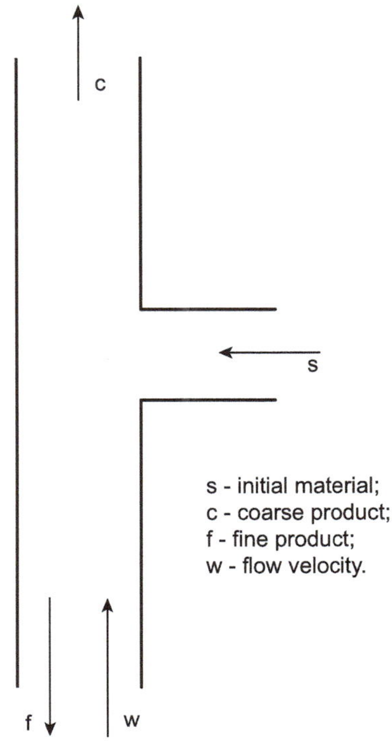

FIGURE 4.1 Diagram of the material distribution in a flow.

s = initial material; c = coarse product; f = fine product; w = flow velocity.

where $F_f(x)$ is fractional extraction upward (fine product); $F_c(x)$ *is* fractional extraction downward (coarse product); γ_f is fine product output; r_f is narrow class content in the fine product; γ_c is coarse product output; r_s is narrow class content in the initial mixture; and r_c is narrow class content in the coarse product. For these ratios, the following is valid:

$$\gamma_f + \gamma_c = 100\%$$

$$r_f\gamma_f + r_c\gamma_c = r_s$$

$$F_f(x) + F_c(x) = 100\%$$

Fractional extraction curve (Figure 4.3) refers to the zone between x_a and x_b points. In the point *a*, $F_f(x) = 100\%$, a transporting regime is realized for a given fraction. In the point *b*, $F_f(x) = 0$ is the falling layer regime for a fraction with the average size *b*. Hence, all conceivable regimes of two-phase flows are realized here.

Experimental curves for various fractions presented in Figure 4.2 show mass transfer character at the velocities within the separation regimes. Hence, we can make the following conclusions. First, extraction of various fractions upward starts at different flow velocities. Second, each fraction has its own velocity at which it is completely extracted, i.e. transportation regime is realized.

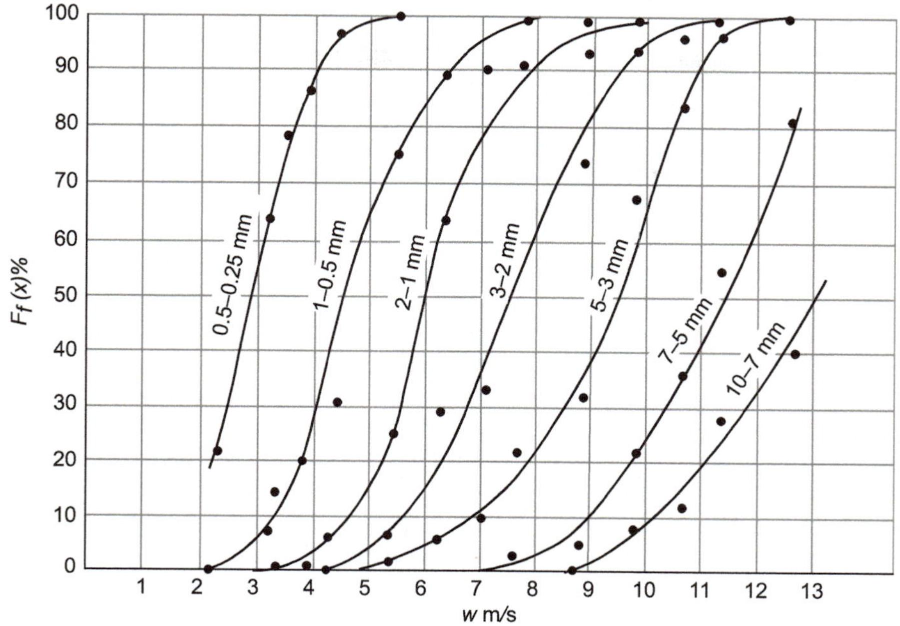

FIGURE 4.2 Fractional extraction of various size classes as a function of air flow velocity in a cascade separator ($z = 4$; $i^* = 1$).

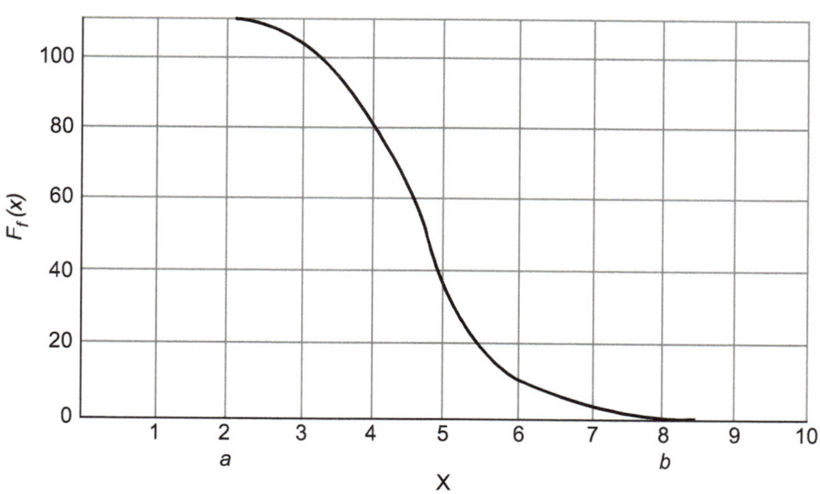

FIGURE 4.3 Separation curve.

Special studies were performed to clarify the effect of composition of the initial bulk material on fractional extraction value. Experiments were carried out for three flow regimes with the velocities 3.15, 4.35 and 5.45 m/s. Each narrow size class was averaged, in its turn, from one experiment to another according to a certain scheme. Its content in the mixture was specified according to the following scheme: 3.3%, 6.7%, 10%, 12.5%, 30%, 50%, 76.7% and 100%. It was assumed that the contents of other classes were uniformly distributed. Numerous experiments were performed, with analogous results. For illustration purposes, the results of one group of experiments are shown in Figure 4.4.

The character of experimentally obtained dependencies points to the fact that, under the conditions being studied, each fraction behaves autonomously, as if there are no other fractions. These results seem somewhat unexpected, since the fact of intense interaction of particles of different fractions in these conditions has been experimentally proved. Particles collisions are definitely random in character. Other random factors also arise; nevertheless, in some way, the invariance of these curves with respect to the initial composition is generated here. Stroboscopic photographs (Figure 4.5) and rapid filming in empty channels and in channels with head pieces have demonstrated a complete disorder in the particle motion and, at the same time, such an unexpected deterministic result.

The invariance of fractional extraction with respect to solid phase composition is of basic importance for the development of a general empirical theory of the process, although the mechanism of such process is absolutely unclear. The same refers to the influence of solid phase concentration on fractional extraction. Concentration increase manifests itself in a dual way. First, obstruction of the solid particles' motion increases. Second, the intensity of their contact interaction increases. Special thorough studies were carried out in an empty vertical channel and in a channel with head pieces. In all cases, the results were qualitatively the same. They can be illustrated by the results of the studies in a hollow pipe (Figure 4.6). A characteristic feature of all fractions without any exception is the presence of a segment parallel to the concentration axis. For the specific case under study, this segment is

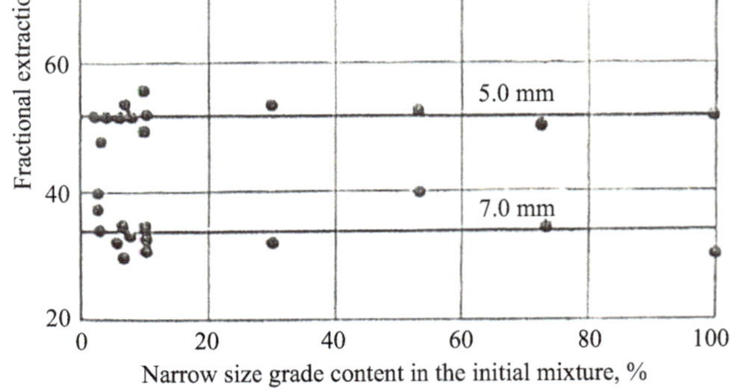

FIGURE 4.4 Fractional extraction of various size classes as a function of their contents in the initial material.

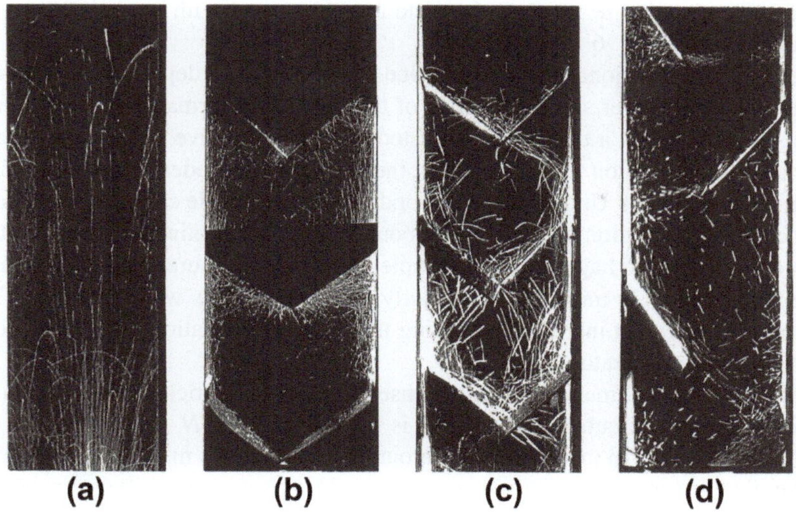

FIGURE 4.5 **Stroboscopic photographs of particle motion in various channels.**

(a) empty; (b) triangular shelves; (c) intersecting shelves at the same level covering a quarter of the cross-section; (d) checkerboard arrangement of shelves covering half of the cross-section.

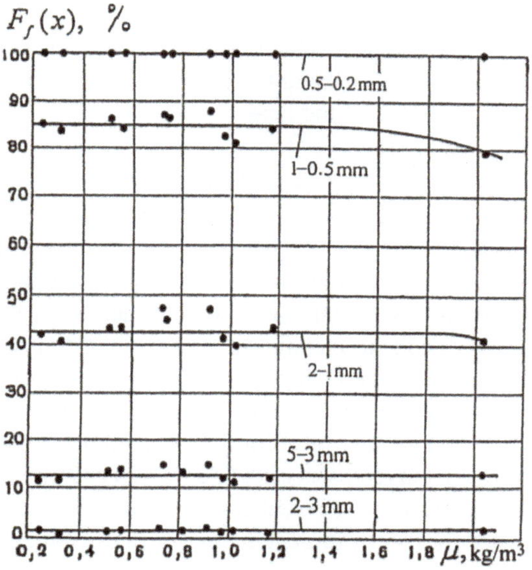

FIGURE 4.6 **Concentration effect on fractional extraction of different size classes.**

restricted by the concentration $\mu \approx 2$ kg/m^3, and for apparatuses with head pieces the region of this regularity extends up to $\mu \approx 6$ kg/m^3.

Within these values, fractional extraction value is practically independent of the material concentration in the flow. However, the mechanism of this regularity formation remains unclear.

Multiple experiments with a thoroughly adjusted flow velocity give the same separation curve at the initial composition variation and changes in the feed by the feeder. This was confirmed by an 11-fold laboratory experiment. Under industrial conditions in a cascade classifier with the productivity of 35 t/h, the curve of separation by the 100-μm boundary was determined in a week after putting the apparatus into operation. It so happened that complete repeated measurements on the same apparatus were possible only after 11 months, and exactly the same curve was obtained. The spread of experimental points somewhat increased, reaching the maximal deviation of 7.3% as compared with 4.4% obtained on a new apparatus.

The process stability is amazing. An immense number of particles take part in the process. The number of particles per cubic meter of air is on the order of $N = 10^{11}$ at the concentration $\mu = 2$ kg/m^3. Air flow rate in the apparatus is around $Q = 20{,}000$ m^3/h. It is a purely chaotic mass process, but the obtained result is strictly deterministic. This phenomenon also remains unexplained.

2. Affinization of fractional extraction curves

The stability of separation curves and their invariance with respect to initial composition and concentration reflects some in-depth laws of two-phase flows. It was of interest to examine them to find at least an empirical approximating dependence. First, a dependence of the following type was plotted:

$$\lg F_f(x) = f\left(\frac{1}{w}\right)$$

This dependence is shown in Figure 4.7. In such a semi-logarithmic interpretation, the dependence is readily rectified in its middle part and deviates to the left when approaching to 100%. This suggested the necessity of analyzing a dependence of the type

$$\lg F_f(x) = f\left(\frac{1}{w^2}\right)$$

shown in Figure 4.8. Here the curves are rectified better. In all cases, the distance from their respective points to the ordinate axis is connected in some way with the inverse proportionality to the average size of each fraction. Multiplying the abscissa by this value gives the dependence shown in Figure 4.9. In this plot, the value $\frac{x}{w^2}$ is multiplied by gravity acceleration g. As a result, a dimensionless complex $\frac{gx}{w^2} = Fr$ is obtained. Its structure reminds one of the Froude criterion in a somewhat modified form, since the size relates to a particle in it, and the velocity relates to the flow. Graphs 4.7, 4.8, and 4.9 are based on the same experiments as graphs in Figure 4.5. In an ordinary coordinate system, these curves give the dependence presented in Figure 4.10. It represents one line with all experimental points reliably fitting to it. Similar dependencies were experimentally revealed for several dozens of various types of separators of different heights and cross-sections with different internal devices. Some of them are shown by way of example in Figure 4.11. It is noteworthy that the correlation coefficient for the indicated approximation for all versions of air separators tested in laboratory conditions varies within the limits from 0.79 to 0.93. This points to a sufficient rigidity of the connection between the parameters under study.

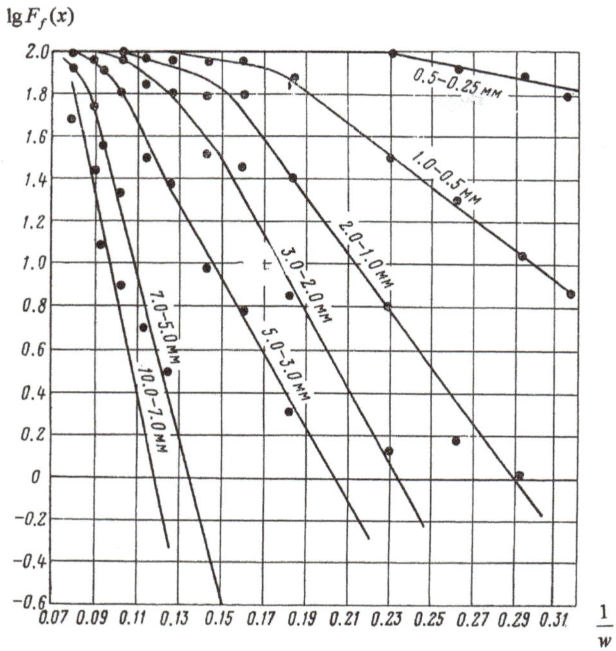

FIGURE 4.7 Fractional separation as a function of the parameter $\frac{1}{w}$ (cascade).

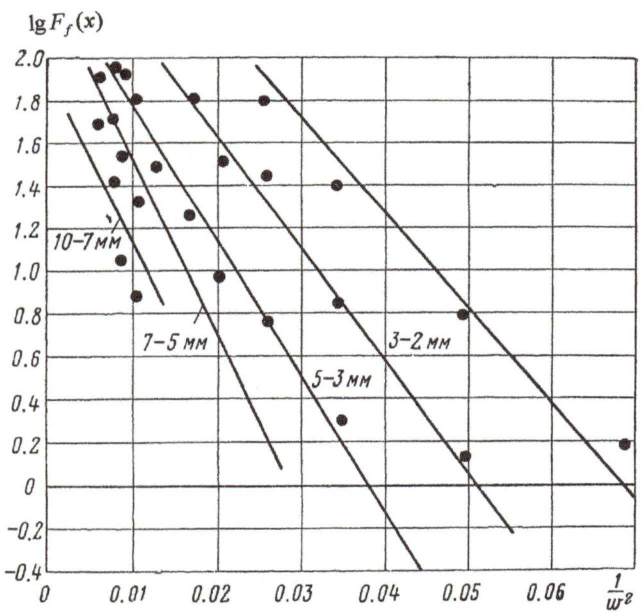

FIGURE 4.8 Fractional separation as a function of the parameter $\frac{1}{w^2}$ (cascade).

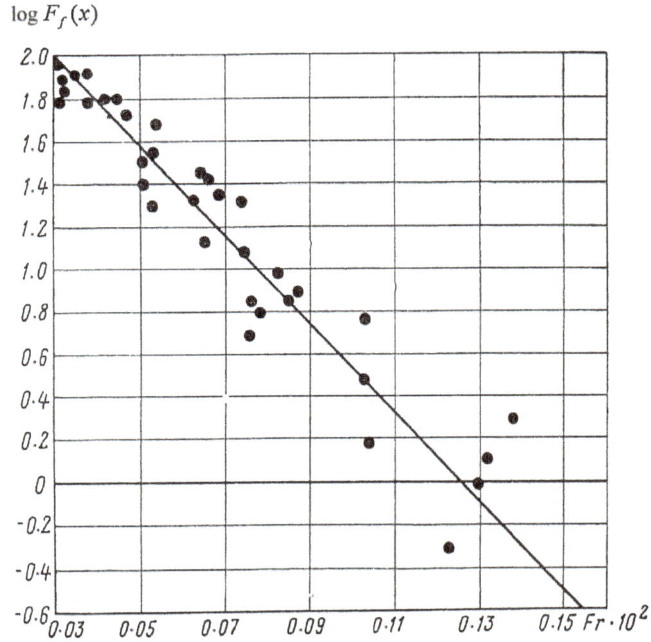

FIGURE 4.9 Fractional separation as a function of the parameter $\frac{x}{w^2}$ (cascade, $z = 4$; $i^* = 1$).

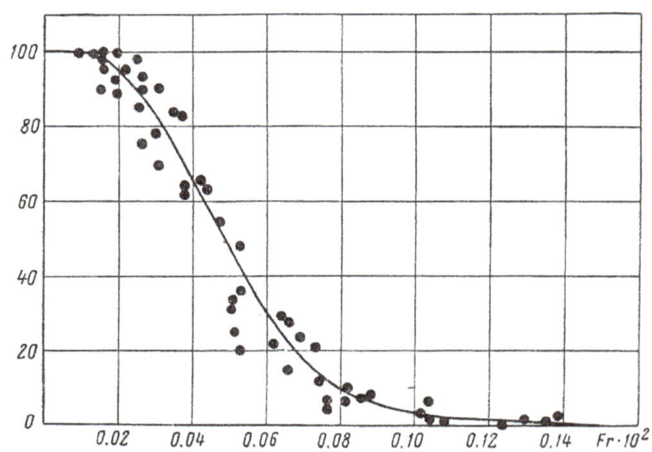

FIGURE 4.10 Fractional separation as a function of the parameter $Fr = \frac{gx}{w^2}$ (cascade, $z = 4$; $i^* = 1$).

Physical meaning of this experimentally established fact is as follows:

1. Separation curves obtained on the same apparatus in turbulent regimes of the ascending medium flow are affine.
2. In spite of visual chaotization of the process, this testifies to the presence of a general physical regularity of the process, according to which the results of fractional separation by various boundary sizes are rigidly connected with medium flow velocities.

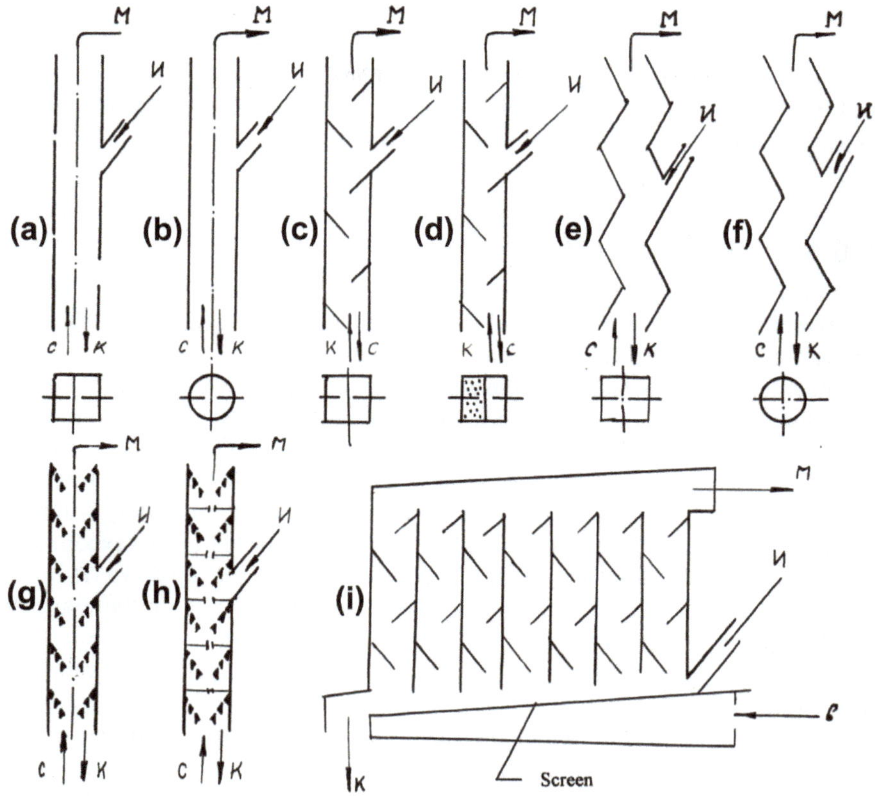

FIGURE 4.11 Various separator designs.

(a) hollow rectangular; (b) hollow round; (c) cascade; (d) cascade with perforated shelves; (e) 'zigzag' rectangular; (f) 'zigzag' round; (g) polycascade rectangular; (h) polycascade round; (i) multi-row with a distributive grate. Notations: s, initial feed; c, coarse product; f, fine product; w, air flow.

3. A curve of $F_f(x) = f(Fr)$ type is invariant not only with respect to the initial composition, but also with respect to fraction sizes and medium flow velocity.
4. Taking into account the fact that this curve is also invariant with respect to material concentration in the flow within a certain working range, we can conclude that it is an unambiguous characteristic of the separator design.

In fact, the position of this curve in the chosen scale is independent of the initial feed composition, fineness of fractions, flow velocity, and productivity variations. However, for separators differing in height, boundary conditions, or configuration, curves of the same type but located differently with respect to coordinate axes are obtained. Hence, they are characteristic of separator design only.

5. An opposite conclusion follows, namely, that separators must be compared on the basis of characteristics of an affine separation curve.

6. The possibility of revealing deterministic regularities for mass transfer in two-phase turbulent flows seems incredible, since even single-phase turbulent flows resist any physical interpretation as yet. At the same time, such determinism is an established fact that has been experimentally confirmed for many years. It is impossible to explain the cause of this phenomenon and its physical aspects from the standpoint of today's knowledge.

3. Unified separation curves

Two-phase flows are frequently used in industry for processes with particles of different densities. The most visual examples are methods of mineral concentration. Ground material is subjected to a flow impact with the purpose of extracting valuable components out of a mixture.

To reveal the influence of the material density on a general regularity, special studies were carried out. For these studies, materials with a broad range of this parameter's variations were selected. Granulometric characteristics and densities of these materials are presented in Table 4.1.

All the materials listed above except potassium salt and cement clinker have particles of rounded shape. We emphasize that the densities of all ore minerals lie almost completely within this region. Two types of cascade classifiers seven stages high ($z = 7$) were selected for the experiments, namely, with the material feed to the second and fourth stages counting from the top down ($i^* = 2$; $i^* = 4$) (Figure 4.11(c)). Separation of each material was performed individually on these apparatuses in a broad range of air flow velocity variation. For each material, graphs of $F_f(x) = f(Fr)$ type were plotted. The obtained curves were affine for each material, but for each density separate curves were obtained even on the same apparatus.

These experiments were completed by the study of separation of mixtures of powders of different densities. A mixture containing 40% of magnetite and 60% of quartzite was prepared. As follows from the table, these two materials have rather close granulometric characteristics, i.e., in this case the effect of the difference in granulometric composition on separation results is practically reduced to zero. All experiments with the mixture were also performed on both separators. Experimental results are

Table 4.1 Bulk Materials Composition in Partial Residues

Material Name	Density	Partial Residues (%) on Sieves with the Mesh X, mm						
Granulated PVC	ρ, kg/m^3	2.5	1.5	1.0	0.75	0.43	0.2	0
	1070	10.1	20.9	28.5	16.3	15.3	7.97	0.93
Potassium salt	1980	13.7	34.1	33.9	5.0	5.3	5.0	3.0
Gypsum macadam	2270	4.1	29.5	23.6	10.3	12.9	11.2	8.4
Quartzite	2675	7.2	27.8	21.2	10.3	15.8	12.8	4.9
Cement clinker	3170	0.2	19.4	25.5	11.7	14.5	11.9	16.8
Magnetite	4350	7.2	26.0	22.8	11.3	13.8	13.3	5.6
Alloy No.1	6210	0.6	10.1	26.8	18.6	25.4	15.2	3.3
Granulated pig-iron	7810	5.3	34.4	39.7	10.7	6.8	2.4	0.7
Alloy No. 2	8650	0.25	0.2	0.15	0.12	0.088	0	–
		12.6	41.8	7.7	7.5	29.7	0.7	–

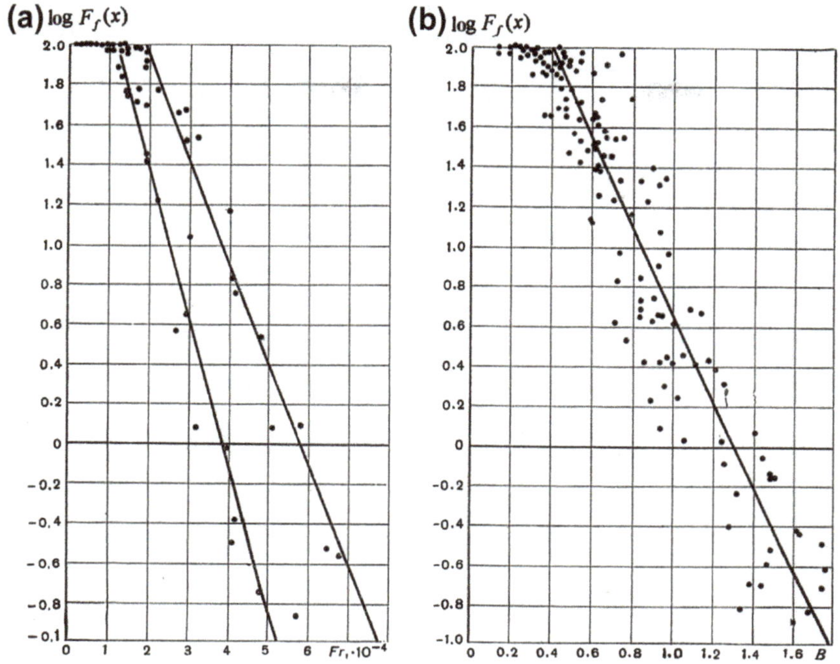

FIGURE 4.12

(a) Nature of the dependence at definite separation of a mixture of materials $F_f(x) = f(Fr)$ for $z=7$; $i^* = 2$.
(b) Nature of the dependence of $F_f(x) = f(B)$ type for the same experimental data.

analogous and correspond to the dependence presented, by way of example, in Figure 4.12 for the apparatus with $z = 7$; $i^* = 2$.

It follows that, in a mixture of materials of different densities, affinization of separation curves takes place separately for each of them. It is of interest that each of these curves exactly coincides with the separation curve obtained for individual material on a specific separator. This fact can be regarded as independence of separation of each component of a mixture from another. We have demonstrated earlier the autonomy of separation of various size classes of the same density at consumed concentrations not exceeding $2 \div 3$ kg/m^3. Now this statement can be extended, since it has turned out that materials of different densities behave autonomously in a flow, as well, despite a sufficiently broad granulometry of particles size in each of them. Attempts were made to find a parameter allowing the affinization of separation curves taking into account different densities. Such a dimensionless parameter was found empirically in the form of a relationship

$$B = Fr \cdot \frac{\rho - \rho_0}{\rho_0} = \frac{gx}{w^2} \cdot \frac{\rho - \rho_0}{\rho_0} \tag{4.2}$$

Application of this parameter provides a reliable affinization of all experimental results for materials of different densities and their mixtures, as shown in Figure 4.13. A somewhat elevated spread of

FIGURE 4.13 Dependence $F_f(x) = f(B)$ for various materials.

(Table 1 cascade shelf separator for $z = 7$; $i^* = 2$.)

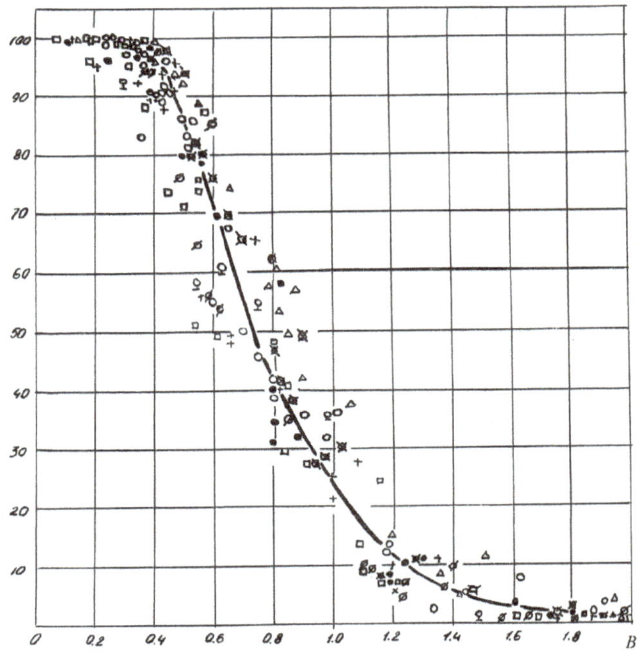

experimental points can be attributed to differences in particle shapes and densities of one and the same material in various size classes.

Thus, an invariant is found that ensures the affinity of fractional extraction curves obtained in a channel of a specific design for materials of various compositions and densities in turbulent flows. The essence of this regularity is that any bulk material is separated equivalently in a specific apparatus according to the same universal curve. The character of this curve is by no means affected by regime parameters, granulometric characteristics, solid phase concentration (within the working range), boundary particle size value, or material density, because, at any ratio of all these parameters, the separation proceeds along the same curve.

A unified curve contains complete information about all possible regimes of two-phase flows in specific conditions. Consider Figure 4.14. For $B < a$ a pneumotransport regime takes place, for $B > b$ a falling layer regime, for B_{50} a regime close to a boiling layer. Mass distribution as such occurs at B parameter variation in the range from a to b. Here fractional extraction varies from 100% to 0. Despite a visual chaos of the process, the unified curve clearly suggests the idea of the absence of randomness in obtaining any value of fractional extraction. Here a practically deterministic regularity takes place; this is rather strange because, above all, the process is characterized by many random factors varying both in space and in time. Hence, the influence of all these factors is leveled in some way under real condition of the process. It is absolutely unclear how it happens, how a unified curve is formed.

The position of the obtained universal dependence in the coordinate system

$$F_f(x) = f(B)$$

reflects only the design of the separating facility realizing the process.

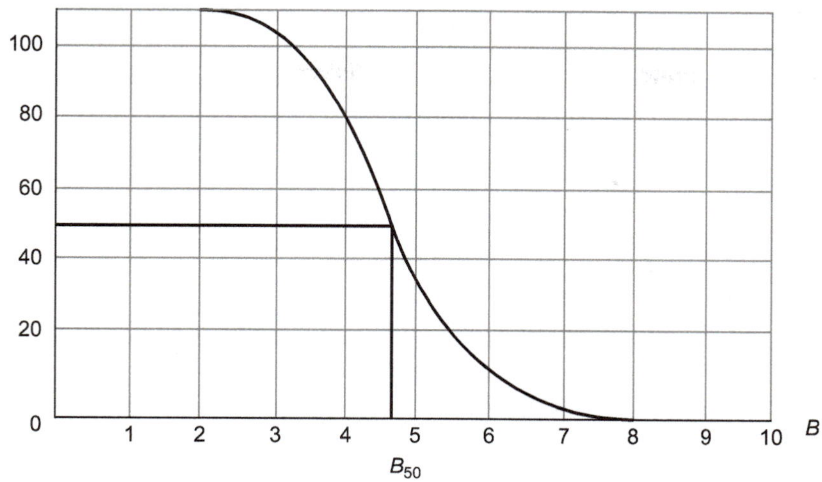

FIGURE 4.14 Separation curve for a narrow size class.

To obtain other curve, one should either introduce some change into the design or replace the entire apparatus. Only these methods allow an impact on the separation process parameters. To understand the physical meaning of the separation curve, we make an attempt to single out some characteristic points on it. For this purpose, examine the character of polyfractional material mass distribution.

We assume that a granulometric characteristic of a certain bulk material composition in partial residues is depicted by ABC curve (Figure 4.15). The graph area restricted by this curve and coordinate axes corresponds, on a certain scale, to the total amount of material. For the sake of convenience, we take it as unity or 100%. The area restricted by ordinates x and $x + \Delta x$ shows the content of a narrow class of x size in this mixture. Assume, in addition, that in the process of mass transfer this mixture is separated into two products. The characteristic of a finer product composition is described by the curve $q(x)$, and that of a coarser one by $n(x)$. Assume the ordinate of the intersection point of these curves as a characteristic size and denote it by x_0.

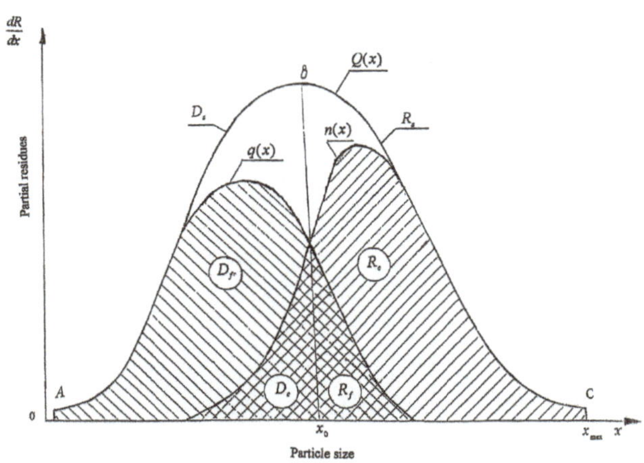

FIGURE 4.15 Bulk material separation into two products.

Introduce the following notations: the area of ABx_0 graph is denoted by D_s, and that of x_0BC graph by R_s. We denote by D_f a part of the area limited by $q(x)$ curve to the left of Bx_0 line, and by R_f the area to the right of this line. We denote by D_c a part of the area $n(x)$ to the left of Bx_0 line, and by R_c the area to the right of this line. Simple equalities arise:

$$\left.\begin{aligned} R_s + D_s &= 1 \\ D_f + D_c &= D_s \\ R_c + R_s &= R_s \\ R_f + D_f &= \gamma_f \\ R_c + D_c &= \gamma_c \end{aligned}\right\} \tag{4.3}$$

where γ_f and γ_c are percentages of products that come out upwards and downwards.

If the technological problem consisted in material separation by size x_0, the separation by Bx_0 would give an ideal result. Unfortunately, nothing ideal occurs in mass processes. In real processes, a part of fine fraction gets into the coarse product, and a part of coarse fraction gets into fine product, as roughly shown in this figure. It is clear from the method of plotting these curves that

$$Q(x_i) = q(x_i) + n(x_i)$$

Dividing both parts of this equality by $Q(x_i)$, we obtain

$$1 = \frac{q(x_i)}{Q(x_i)} + \frac{n(x_i)}{Q(x_i)}$$

This is analogous to

$$1 = F_f(x_i) + F_c(x_i) \tag{4.4}$$

where $F_f(x_i)$ is narrow class extraction into the fine product; $F_c(x_i)$ is extraction of the same class into the coarse product. Note also that D_f and R_c are target separation products. The higher their percentage, the higher is the quality of the process organization. D_c and R_f are contaminating components of the products.

To define conditions for a maximal difference in separation products, it is necessary to optimize the relation

$$E_I = D_f + R_c$$

or to minimize

$$E_{II} = R_f + D_c$$

The optimality condition according to these dependencies is

$$\frac{dE_I}{dx} = 0; \quad \frac{dE_{II}}{dx} = 0.$$

We expand the first condition

$$\frac{dE_I}{dx} = \frac{dD_f}{dx} + \frac{dR_c}{dx} = \frac{d\int\limits_0^x q(x)dx}{dx} + \frac{d\int\limits_x^{x_{\max}} n(x)dx}{dc} = 0$$

As known, a derivative of a definite integral with a variable upper limit and a constant lower limit equals the integrand in the upper limit point. Hence, we obtain

$$q(x) - n(x) = 0,$$

i.e.,

$$q(x) = n(x)$$

The expansion of the second condition gives a similar result. It means that in the optimum point the narrow class is divided in half, i.e. the following relation corresponds to the optimality condition:

$$F_f(x) = F_c(x) = 50\% \tag{4.5}$$

Thus, three characteristic points are obtained on the separation curve:

1. $F_f(x) = 100\%$—pneumotransport condition;
2. $F_f(x) = 0\%$—falling layer (bed) condition;
3. $F_f(x) = 50\%$ defines optimal separation condition. The rest of the points in this curve correspond to the parameters of two-phase flow ensuring mass distribution of the solid phase between two products.

At the same time, parameters characteristic of a single particle, such as the Froude criterion (*Fr*) or a dimensionless ratio B are invariants leading to the affinization of separation curves.

As known, particle hovering velocity is defined by the relation

$$w_0 = \sqrt{\frac{4gx(\rho - \rho_0)}{3\lambda_0 \rho_0}}$$

where λ_0 is particle resistance under hovering conditions.

We transform this expression:

$$\frac{3\lambda_0}{4} = \frac{gx(\rho - \rho_0)}{w_0^2 \cdot \rho_0} \tag{4.6}$$

It is experimentally established that, in a hollow apparatus at the flow velocity equal to hovering velocity of narrow class particles, $F_f(x) = 50\%$. It follows from the dependence (4.6) that

$$\frac{3\lambda_0}{4} = B_{50} \tag{4.7}$$

This means that, for one apparatus, $B_{50} = const$ acquires the meaning of drag coefficient of a solitary particle. In this connection, a question arises of determining at the same physical level the values of other parameters B ensuring equal extractability of various classes, for example, by 10%, 20%, 40%, 70%, 95%, etc.

As known, a particle motion velocity in an idealized flow is

$$v = w - w_0 \tag{4.8}$$

Hence, in a general case, flow velocity can be represented as a sum of particle velocity and its hovering velocity, i.e.

$$w = v + w_0$$

A general expression for the parameter B in this case can be written as

$$B = \frac{gx}{w^2}\frac{(\rho - \rho_0)}{\rho_0} = \frac{gx(\rho - \rho_0)}{(v + w_0)^2 \rho_0} = \frac{3}{4}\lambda \tag{4.9}$$

Here λ is drag coefficient of a particle for a specified value $F_f(x) = const.$

According to Eq. (4.7),

$$B_{50} = \frac{gx}{w_0^2}\frac{(\rho - \rho_0)}{\rho_0} = \frac{3}{4}\lambda_0 \tag{4.10}$$

We divide Eq. (4.9) by Eq. (4.10):

$$\frac{B}{B_{50}} = \frac{\lambda}{\lambda_0} = \frac{w_0^2}{w^2} \tag{4.11}$$

Hence, the ratio of flow velocity to hovering velocity of narrow class particles predetermines the equal extractability value of various fractions.

If the character of the dependence $F_f(x) = f(B)$ is universal for turbulent flows only, one can expect that a dependence of $F_f(x) = \varphi\left(\frac{B}{B_{50}}\right)$ type acquires a universal character for any regimes of medium motion.

4. Generalizing empirical invariant

To determine the generalizing invariant, it was necessary to carry out experiments in the velocity range comprising turbulent, transient, and laminar motion regimes. Clearly, for such an experiment, a low-density material with fine particles is necessary. For this purpose, aluminum powder with the density 2700 kg/m was chosen, which is used for making paints. The experiments were performed on a cascade classifier (Figure 4.11(c)) of nine stages with material feed into the apparatus at the middle stage ($z = 9$; $i^* = 5$). Granulometric composition of the aluminum powder in partial residues is given in Table 4.2.

Since all regularities for these particles can drastically differ from those for coarser ones, first of all, the effect of material concentration on separation results was determined. Three sets of experiments were performed at the flow velocity $w = 0.53$ m/s and consumed solid phase concentrations equal to 2.75; 6.0 and 14.3 kg/m^3. It turned out that, in this range of concentration variation, its effect is insignificant. This is evident from the graph in Figure 4.15, where all experimental results give one universal curve. The spread of experimental points is not large and can be attributed to errors due to the difficulties in determining the granulometric composition of particles of the size range under study.

In this concentration range, experiments were carried out on the same apparatus at various air flow velocities varying within the range from 0.29 m/s to 1.46 m/s. Some experiments in this set were performed three times. On the same apparatus, experiments on quartzite powder separation with a close density of $\rho = 2670$ kg/m^3, but with elevated (in comparison with aluminum powder) particle size from 0.1 to 3.0 mm, were performed. Experiments on quartzite were performed at air flow velocities of 4.7, 5.57, 6.67, 7.3 and 7.89 m/s.

Table 4.2 Granulometric composition of the aluminum powder

Boundary Size, *x*, mm																		
0.4	0.315	0.2	0.16	0.125	0.1	0.08	0.063	0.05	0.045	0.04	0.035	0.03	0.025	0.02	0.015	0.01	0.005	0
Partial Residues, *z*, %																		
6.5	6.9	14.3	7.8	6.5	6.7	7.8	8.2	3.5	4.8	5.2	4.2	4.8	4.2	3.0	3.8	1.0	0.5	0.3

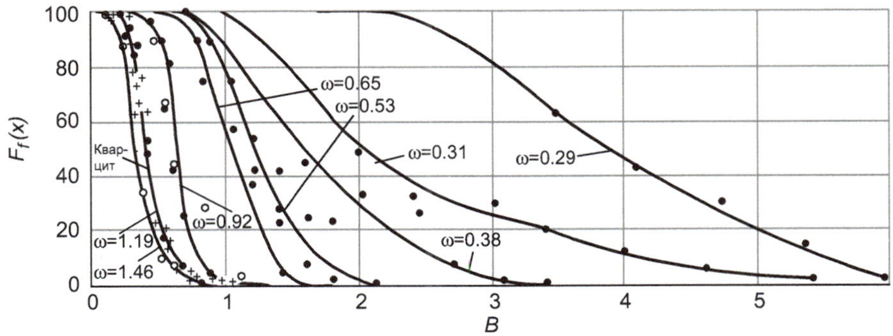

FIGURE 4.16 Dependence $F_f(x) = f(B)$ for the separation of extremely fine powders.

All experiments were processed using the $F_f(x) = f(B)$ method and plotted in one graph (Figure 4.16). All experimental results with quartzite ensure stable affinization. Practically all experimental points obtained in aluminum powder fractionating at elevated velocities equal to 1.19 and 1.46 m/s fit this curve too.

At lower flow velocities, curves of similar shape are obtained; however, they diverge, and the lower the flow velocity, the more intensely they diverge. Here we deal with a new situation. To analyze it, on the basis of experimental data, for each flow velocity a dependence of $F_f(x) = f(x)$ type was plotted, and for each flow velocity the values of x_{50}, and then of B_{50} were determined. The results of these estimations are summarized in Table 4.3.

On the basis of these data, the Reynolds number value Re_{50} (4th line) is calculated. Here the dependence of $B_{50} = f(\mathrm{Re}_{50})$ type is of greatest interest. It is presented in Figure 4.17. Keeping in mind that B_{50} corresponds to drag coefficient λ_{50}, one can easily see a complete analogy of the obtained dependence to the Rayleigh curve for a single particle of the $\lambda = f(\mathrm{Re})$ type.

This experimental dependence can explain a lot. In a horizontal segment, to the right of the $\lg \mathrm{Re}_{50} = 1$ value, $B_{50} = const$. This ensures unambiguous affinization of separation curves with respect to the criterion Fr or parameter B. This is the region of turbulent flow. At a transition to a laminar flow, the affinization with respect to these parameters is violated. On this specific apparatus, a transition from one regime to another is realized at $R_{50} \approx 4.76$, which corresponds to the boundary size of 0.056 mm at $\rho = 2700\ \mathrm{kg/m^3}$.

Hence, the affinization and universality of separation curves take place for separation boundaries above 56 μm, while below this value the regularity is violated. Obviously, for materials of other densities, this boundary will be different. Taking into account the generalizing nature of the parameter

Table 4.3 Main Parameters of Aluminum Powder Separation on $z = 9$; $i^* = 5$ Apparatus

Flow velocity, w m/s	1.46	1.19	0.92	0.65	0.53	0.38	0.31	0.29
Granul. size, x_{50}, mm	0.078	0.056	0.05	0.043	0.032	0.020	0.015	0.013
B_{50}	0.035	0.41	0.65	1.1	1.23	1.6	2.0	3.9
Re_{50}	8.14	4.76	3.28	1.99	1.21	0.54	0.33	0.26

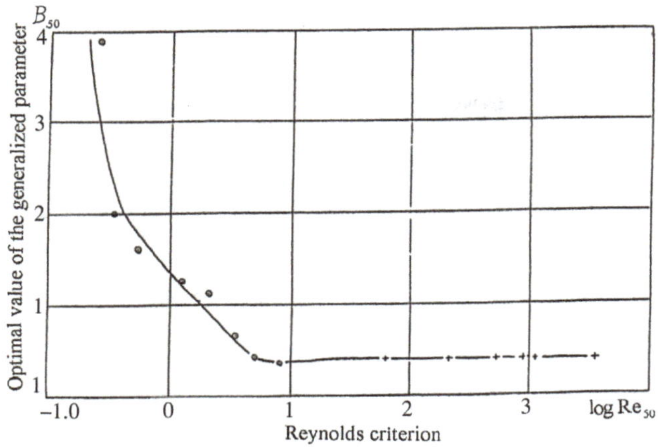

FIGURE 4.17 Dependence $B_{50} = f(Re_{50})$.

B, we can determine this boundary for a material of any density. Therefore, an idea arises that it is necessary to introduce a new parameter of particle size into the practice of two-phase flow estimation, namely, a hydraulic size taking into account the densities of solid particles and moving medium. It can be determined as

$$x_g = x\frac{\rho - \rho_0}{\rho_0}$$

Using this parameter, we can considerably simplify the previously found parameter writing, for example

$$Fr = \frac{gx_g}{w^2}; \quad \mathrm{Re} = \frac{x_g w}{\nu} \quad \text{etc.}$$

In such an interpretation, dimensionless similarity criteria and invariants acquire a more generalized meaning.

Powders similar to aluminum powder acquire, in aqueous medium, a region of turbulent overflow for particles of the following size:

$$0.057\frac{2700 - 1.29}{1.29} = x\frac{2700 - 1000}{1000} \quad x > 70\text{mm}$$

For a material with $\rho = 3000$ kg/m^3 in an air flow, turbulence boundary starts from 45 μm, reaching $x > 52$ mm in aqueous medium. Hence, regularities similar to those for very fine particles in air flows are also valid in water flows.

However, here it is also possible to reduce all experimental values given in Figure 4.16 to one affine curve. This became possible by accomplishing the following procedure. For each curve, a B_{50} value (Table 4.3) was determined, and then a dependence of the following type was plotted:

$$F_f(x) = f\left(\frac{B}{B_{50}}\right)$$

FIGURE 4.18 Affine dependence $F_f(x) = f\left(\frac{B}{B_{50}}\right)$.

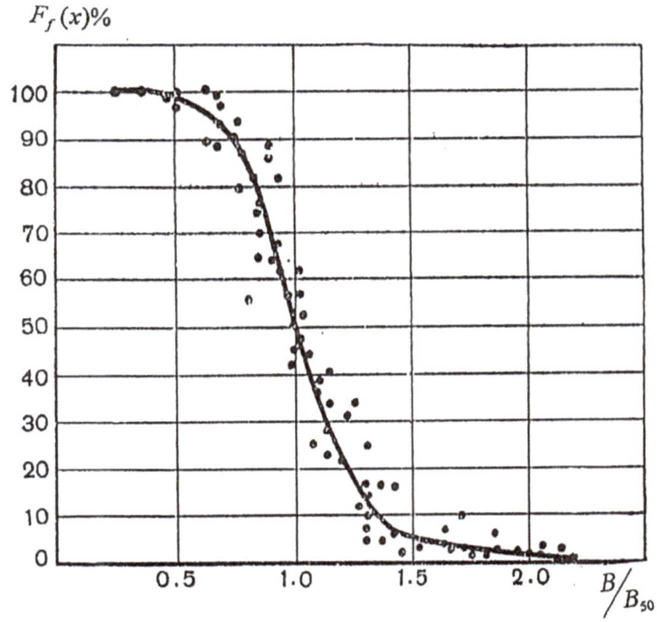

A curve obtained in this way including all (without any exceptions) experimental values from Figure 4.16 is given in Figure 4.18. As the figure shows, here we obtain a generalized affinization of experimental separation curves for any range of particle size and any media. In ordinary conditions, this range starts from particle size on the order of 5–6 μm. Under special conditions it can be reduced to less than 1 μm.

CHAPTER

Entropy of Mixture Composition

5

CHAPTER OUTLINE

1. Invariants for granulometric composition 79
2. Uncertainty of mixture composition 83

1. Invariants for granulometric composition

In bulk materials separation, as well as in other separation processes, we deal, as a rule, with multicomponent compositions. As a result, initial mixtures are separated into two or more materials. Besides, each of these materials can contain various components of the initial mixture in arbitrary ratios. Here we are facing a complicated problem of unambiguous estimation of a mixture composition, no matter how complicated it is. Without such estimation, it is difficult to perform separation processes, because the plurality of parameters characterizing final compositions of products does not allow the optimization of this process. Many attempts were made to develop quality criteria for binary mixtures separation. One of the most widespread is Hancock's method, but it has essential defects even in the case of binary separation. This method, like many similar methods, cannot estimate separation if the number of products obtained from the initial mixture exceeds two. The physical meaning of the notion of "quality criterion" or "efficiency" in the case of binary separation remains necessary, but becomes insufficient at the analysis of a multicomponent mixture separation even into two products. At a greater number of final products, even ideal extraction of one of components of a polyfractional mixture does not testify to a 100% efficiency of the entire process, since, at the same time, other components can be nonideally separated.

Apparently, we will be able to evaluate the quality of n-component mixture separation into m products after realizing the following ideas.

It is easy to express unambiguously a binary mixture composition, since the content of one component determines the content of another because their sum equals unity or 100%. It means that a binary mixture has one degree of freedom. Evidently, $n-$ component mixture has $(n-1)$ degrees of freedom, and the determination of its components in a usual way is never unambiguous.

Thus, a method is needed that will allow an unambiguous (numerical) characterization of the general composition of a multicomponent mixture. This method must take into account the percentage of each component, both in the initial mixture and in separation products.

Entropic Invariants of Two-Phase Flows. http://dx.doi.org/10.1016/B978-0-12-801458-5.00005-8

Compositions of any mixtures can be interpreted formally and to the point by probabilistic characteristics.

A set of balls of various sizes or colours is used in discussions for implicit clarification of the notion of probability, since it is extremely difficult to directly define such a fundamental notion.

Probability is a partly subjective, quantitative expression of expectation of the result of some random event, experiment, or test. It can be measured only if the event or experiment can be repeated several times. A random value usually implies a parameter (as a rule, numerical), the value of which is determined by the outcome of a random event. As an example, throwing a coin or a six-sided die is usually cited. If a coin is thrown many times, the probabilities of getting heads or tails is approximately the same, i.e. $P = \frac{1}{2}$, since a coin provides only two outcomes. The probability of getting each side of a die is $P = \frac{1}{6}$, since it provides six outcomes.

If some outcome A is obtained in any test, then its probability equals $P(A) = 1$. If, however, the outcome is sometimes different, say, B, then the probability of obtaining A is

$$P(A) = 1 - P(B)$$

Outcomes of sequential tests are distributed at random, if there is no regularity in their sequential appearance. The only possible prediction of the outcome of the next test is reduced to the revealing of probabilities of other possible outcomes. In this respect, the following experiment is characteristic.

Imagine an opaque container with many balls of different colours. If one takes balls at random out of this container, the probability of the appearance of a definite colour is determined, in a long-term test, by the percentage of this colour in the ball mixture. By enumerating in this way the probabilities of getting balls of all the colours, we can obtain their colour distribution function. If the balls differ in size, such an experiment gives their size distribution function or their granulometric characteristic. Thus, a distribution function implies the enumeration of probabilities of all possible outcomes of an event. If, in the example under study, there are N multicoloured balls, the sum of all extracted balls at each step of the experiment equals an integer from 0 to N. The sum of all the probabilities covers all possibilities and, hence, is a certainty, i.e.

$$\sum_N P_N = 1$$

It is clear from this example that granulometric composition can be conveniently interpreted through the probability. If we make a mark at random on the line of particles dimensions with the length L, which has equal probability to be located in any part of this segment, the probability of finding the mark in the region between x and $x + dx$ equals

$$P = \frac{dx}{L}$$

Since the contents of particles of various sizes in a mixture are not the same, the probability of the presence of particles of each size is determined by a weight function, which may be called a function of size distribution of particles (Figure 5.1). Taking this into account, we can write for the dx segment

$$dP(x) = \frac{dR}{dx} dx \tag{5.1}$$

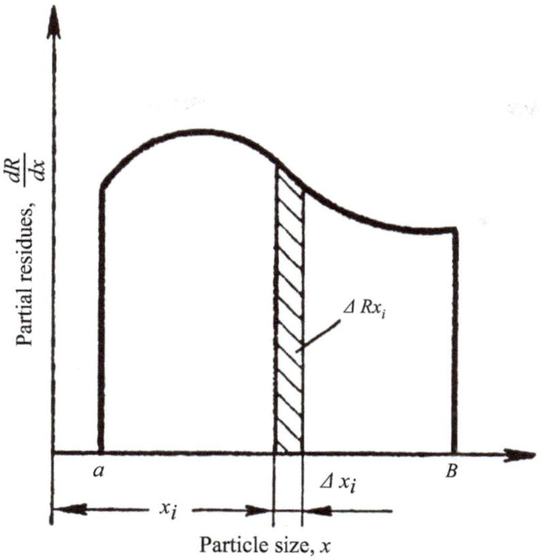

FIGURE 5.1 Granulometric characteristic of polyfractional mixture of bulk material.

The value $dP(x)$ determines the probability of finding particles of a definite size within a dx interval, and includes both the weight function and parameters of uniserial distribution. Since the value dx is located with certainty between 0 and L, the following integral relation is valid:

$$\int_0^L P(x)dx = 1$$

If the equation of the curve in Figure 5.1 is known, the distribution function can be introduced instead of the weight function. The average value or statistical expectancy of a random value $R(x)$, if it is specified continuously, is its integral over the probabilistic space where it is specified:

$$\overline{R(x)} = < R(x) > = \int_{x_a}^{x_b} \frac{dR}{dx} dx \tag{5.2}$$

If a random value is specified discretely, its average value is determined from a great number of its independent realizations:

$$< R(x) >= \sum_n \frac{R(x)}{x_b - x_a}$$

Besides the size, the distribution presented in Figure 5.1 can be characterized by different densities of these particles (Figure 5.2). In this case, the interrelation between the probabilities is expressed

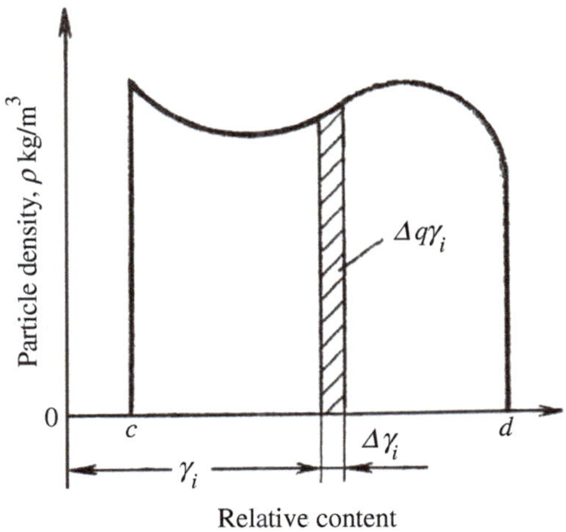

FIGURE 5.2 Bulk material distribution in particles densities.

through a conditional probability $P(A/B)$. The event A takes place in the test, if the event B takes place. In this case,

$$P(A/B) = P\left(\frac{A}{B}\right) \cdot P(B) = P\left(\frac{B}{A}\right) \cdot P(A),$$

and in the case of their sum, $P(A+B) = P(A) + P(B) - P(AB)$.

In the analysis of density distribution, the normalization condition is also valid:

$$N = \int_0^L f(\rho)d\rho,$$

and the integral relation of the contents probabilities gives

$$\int_0^L P(\rho)d\rho = 1$$

Unfortunately, no successful mathematical approximations of $f(x)$ and $f(\rho)$ type have yet been found for bulk materials. One has to use tables of experimental analysis of powder composition by particle size and density in computations. Nevertheless, all relations determined in this section are extremely useful, and will be used in the analysis of multicomponent compositions as well as for further derivations.

Multicomponent mixtures most often occur in nature. Their processing sometimes requires one component extraction out of the mixture, and more rarely simultaneous extraction of two or more components. In this case, the objective of the process is to extract each target component separately and, wherever possible, maximally purified from other components. By way of example, we can mention polymetal ores, coal-containing particles of different densities, or other minerals containing, besides the target component, considerable amounts of waste rock.

A binary mixture can be described by

$$D_s + R_s = 1$$

where D_s is the amount of particles finer than x or lighter than ρ in the mixture, and R_s is the amount of particles coarser than x or heavier than ρ in the mixture. The sum of probabilities of all components of any mixture is unity, and therefore determining the probabilities of some components only does not give clear information about the contents of other components. Thus, an uncertainty arises.

It is perfectly clear that the effect of separation can be determined only by comparing compositions of the initial and final products. Clearly, this problem cannot be unambiguously solved using probability only.

A different approach is needed based on an unambiguous characteristic of any arbitrarily complicated mixture. This issue is of basic and phenomenological character, since, besides particle mixtures, it should be suitable for the evaluation of any possible mixtures (e.g. mixtures of gases, liquids, complex liquid-and-gas mixtures), for isotopes separation, etc.

2. Uncertainty of mixture composition

As already mentioned, any mixture composition can be interpreted through probability. However, as already shown, it is impossible to use it for evaluating multicomponent mixtures. Therefore, we resort to another characteristic of a random value – a number of outcomes or uncertainty measure introduced by Hartley for a totally different problem. The meaning of this parameter is the following.

If some random value x_i has k equiprobable outcomes (for a coin $k = 2$; for a die $k = 6$), the number of outcomes and the probability are interconnected by a relationship

$$P(x_i) = \frac{1}{k}$$

It is clear, even at an intuitive level, that any probabilistic event implies some uncertainty, which must be a function of the number of outcomes $f(k)$.

We will formally define the uncertainty inherent to some random event. It will allow us to determine a parameter of state with a countable number of outputs. We take into account that the uncertainty in the onset of a specified event that existed before the onset of the event should coincide with the amount of information obtained in case this event actually occurs. Hence, the amount of information contained in a random event is connected with an unambiguous quantitative measure of uncertainty.

We can say that uncertainty is a function of events depending on their probabilities only and satisfying the following conditions:

1. An event with the probability equal to unity has a zero uncertainty, $f(1) = 0$.
2. If one event has a smaller probability than another, the uncertainty of the first event is greater than the uncertainty of the second.
3. The uncertainty of simultaneous onset of two independent events equals a sum of their uncertainties.

We try, insofar as possible, to express these conditions through some known functions. Since such a function should depend on the event probability only, it should be determined in the segment [0, 1] and also satisfy the stated conditions. It should monotonically decrease in this segment and satisfy a functional equation

$$k(xy) = k(x) + k(y) \tag{5.3}$$

All three stated conditions satisfy a logarithmic dependence, i.e.

$$f(k) = \begin{cases} c \cdot \log k = H, & 0 < P \le 1 \\ \infty & P = 0 \end{cases} \tag{5.4}$$

where c is a proportionality factor, H is the uncertainty of random event, and *log k* is a value determined to within a constant.

The dependence (5.4) possesses all properties of uncertainty; moreover, it is the only function satisfying all three conditions stated above.

To proceed to computations, it is necessary to define some parameters. The dependence (5.4) as it is corresponds to entropy formula introduced by Boltzmann for a thermodynamic system. Therefore, the parameter $f(k)$ evaluating the composition heterogeneity can be called entropy of mixture composition. Its physical meaning is identical to that of entropy, i.e. $f(k) = H$ for the case under consideration.

Entropy of a random mass event is statistical expectation of the totality of uncertainties of this event, i.e.

$$H = \int P(f)k(f) = -c\sum P(A)\log A \tag{5.5}$$

Here the notion of entropy is accepted as an average quantity of uncertainty in a random mass process. A random process is an abstract mathematical notion used to describe random phenomena depending on a certain parameter that obeys probabilistic laws. The probabilistic law governing the development of a random process is specified by common probabilistic distributions of random magnitudes determining this process.

Taking into account these preliminary notes, we first examine a binary mixture. Let the number of fine particles in it beN_1, and that of coarse particles N_2. The total number of particles in the mixture is

$$G = N_1 + N_2$$

By definition, the probability of a fine particle is

$$P_1 = \frac{N_1}{G},$$

and the probability of a coarse one is

$$P_2 = \frac{N_2}{G}$$

According to statistical physics, the number of outcomes at random extraction of coarse and fine particles out of a mixture is

$$k = \frac{G!}{N_1!N_2!}$$

where k is the number of possible states of the system under study.

According to (5.4), the uncertainty can be expressed by a relation

$$H(k) = c \log k = c(\log G! - \log N_1! - \log N_2!) \tag{5.6}$$

If the values N_1; N_2; G are high enough, logarithms can be estimated using an approximate Stirling formula

$$\log A! = A(\log A - 1) \tag{5.7}$$

Taking this into account, Eq. (5.6) is transformed into

$$H(k) = c[G(\log G - 1) - N_1(\log N_1 - 1) - N_2(\log N_2 - 1)]$$

We factorize G and open the brackets

$$H(k) = -cG\left(\frac{N_1}{G}\log\frac{N_1}{G} + \frac{N_2}{G}\log\frac{N_2}{G}\right)$$

According to the definition of probability, we can write it as follows:

$$H(k) = -cG(P_1 \log P_1 + P_2 \log P_2) \tag{5.8}$$

The value c in Eq. (5.8) is arbitrary; we assume that $c = 1$. Then the uncertainty of a specific system per one element of this system is

$$H(x) = \frac{H(k)}{G} = -(P_1 \log P_1 + P_2 \log P_2) \tag{5.9}$$

The relation (5.9) expresses uncertainty or entropy determined for one particle in a system of G particles (Figure 5.3).

We define an expression of uncertainty for a polyfractional mixture. We assume that it consists of n fractions. We denote particles contents in each fraction by the line $N_1;N_2;N_3;\ldots N_i\ldots;N_n$. The total number of particles is

$$G = \sum_n N_i$$

Then the probability of the presence of a particle of each class is

$$P_i = \frac{N_i}{G}$$

By definition,

$$\sum_n P_i = \sum_n \frac{N_i}{G} = 1$$

FIGURE 5.3 Entropy dependence on the binary mixture composition ratio.

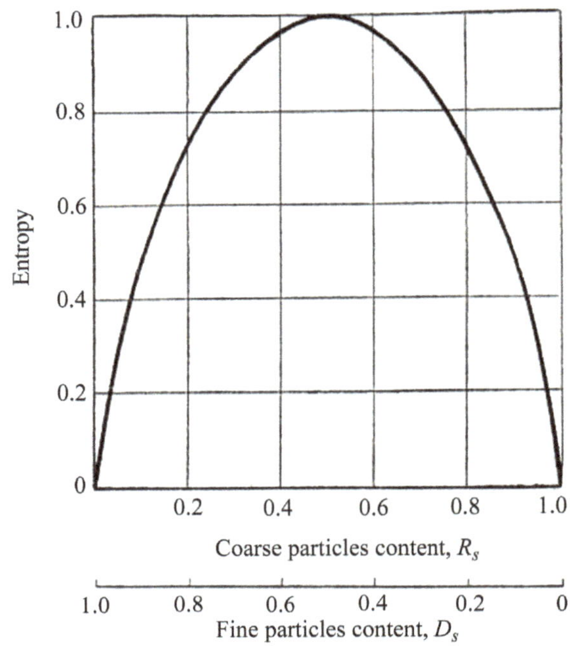

The total number of outcomes for this mixture of particles is

$$k = \frac{G!}{\prod_{i=1}^{n} N_1!}$$

Hence,

$$H(k) = c \log m = c\left[G(\log G - 1) - \sum_{i=1}^{n} N_i(\log N_i - 1)\right]$$

In compliance with the previous derivation, we obtain

$$H(k) = -cG \sum_{i=1}^{n} P_i \log P_i$$

For $c = 1$, the uncertainty of composition of a specific multicomponent mixture per one particle is

$$H(x) = -\sum_{u=1}^{n} P_1 \log P_1$$

We examine a more complicated version. Assume that, in the previous case, each fraction of particles has its own internal density distribution. Then the content of each class by size and density can be expressed by a matrix

$$\begin{vmatrix} M_{11} & M_{21} & M_{13} & \cdots\cdots & M_{1i} & \cdots\cdots & M_{1m} \\ M_{21} & M_{22} & M_{23} & \cdots\cdots & M_{2i} & \cdots\cdots & M_{2m} \\ \vdots & & & & & & \\ M_{j1} & M_{j2} & M_{j3} & \cdots\cdots & M_{ji} & \cdots\cdots & M_{jm} \\ \vdots & & & & & & \\ M_{n1} & M_{n2} & M_{n3} & \cdots\cdots & M_{ni} & \cdots\cdots & M_{nm} \end{vmatrix}$$

Clearly, for each line

$$\sum_{j=1}^{m} M_{jn} = N_j,$$

and for the entire matrix

$$\sum_{j=1}^{m} \sum_{i=1}^{n} \frac{M_{ji}}{G} = 1,$$

which is equivalent to

$$\sum_{j=1}^{m} \sum_{i=1}^{n} P_{ji} = 1 \tag{5.10}$$

The total number of outcomes for said mixture amounts to

$$k = \frac{G!}{\prod_{i=1}^{n} \prod_{j=1}^{m} M_{ji}}$$

Hence,

$$H(k) = c \log k = c\left[G(\log G - 1) - \sum_{i=1}^{n} \sum_{j=1}^{m} M_{ji}(\log M_{ji} - 1)\right]$$

As a result, we obtain for this case

$$H(x) = -\sum_{i=1}^{n} \sum_{j=1}^{m} P_{ij} \log P_{ij} \tag{5.11}$$

The composition uncertainty unambiguously characterizes its heterogeneity. If a mixture consists of identical particles, its uncertainty is zero, and there is no heterogeneity. Thus, it is possible to evaluate mixtures of various composition, and even very complicated ones, by one number. According to this method, different compositions acquire different but unambiguous evaluations. This can become a key to the estimation of changes in mixture compositions in the course of separation processes or similar procedures leading to granulometric composition changes, for example, grinding.

Formulas of composition heterogeneity comprise a logarithm. In principle, it can have any base. The values of logarithm of certain quantities with different bases differ in a constant, because

$$\log_y x = \frac{\lg_{10} x}{\lg_{10} y}$$

Therefore, we should choose the logarithm base value for mixtures, which can make our calculations comparable.

We consider a binary mixture first. An unambiguous heterogeneity characteristic for it can be expressed as follows:

$$H_s(x) = \sum_{1}^{2} P_i \log P_i \tag{5.12}$$

Referring to Figure 4.15, the same expression can be rewritten as

$$H_s(x) = -(R_s \log R_s + D_s \log D_s) \tag{5.13}$$

We determine the maximal possible value of the composition heterogeneity on the basis of habitual estimations, taking it as unity, i.e. $H_s(x)_{max} = 1$. As known, $R_s + D_s = 1$. Therefore, we can write

$$H_s(x) = -R_s \log R_s - (1 - R_s)\log(1 - R_s)$$

We take a derivative of the obtained expression and set it equal to zero:

$$\frac{dH_s(x)}{dR_s} = -[1 + \log R_s - 1 - \log(1 - R_s)] = 0$$

Hence,

$$\log R_s = \log(1 - R_s), \text{ i.e.}$$
$$R_s = 0,5$$

Binary mixture heterogeneity is maximal at equal contents of both components. We substitute the obtained value of R_s into (5.12):

$$H_s(x)_{\max} = -\frac{1}{2}\log\frac{1}{2} - \frac{1}{2}\log\frac{1}{2} = -\log\frac{1}{2} = \log 2$$

To satisfy the formulated requirement, it is necessary to take 2 as the logarithm base in this case. Thus,

$$H_s(x)_{\max} = \log_2 2 = 1$$

The dependence of $H_s(x)$ on the ratio between D_s and R_s is shown in Figure 5.3. For $R_s = 0$, $D_s = 1$, and for $D_s = 0$, $R_s = 1$. The mixture composition in these limiting points is homogeneous. Therefore, the uncertainty of such compositions is zero.

Consider now a three-component mixture. For this mixture, the following relation is valid:

$$P_1 + P_2 + P_3 = 1$$

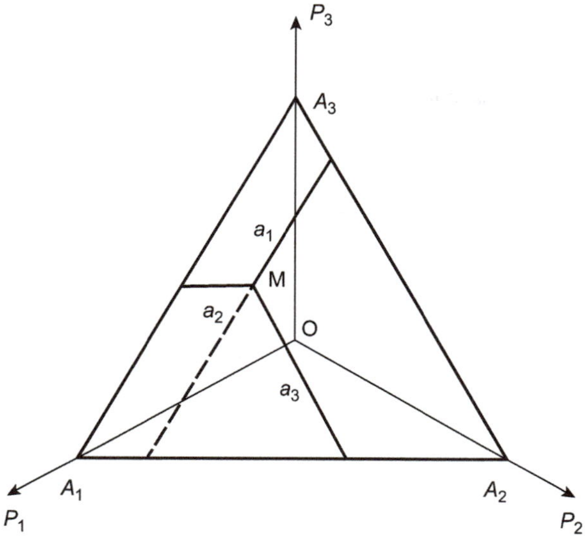

FIGURE 5.4 Planar diagram of a three-component mixture.

According to Eq. (5.9), in this case

$$H_s(x) = -(P_1 \log P_1 + P_2 \log P_2 + P_3 \log P_3) \tag{5.14}$$

First, we plot a diagram for a three-component mixture and draw three coordinate axes in a plane at the angle of 120° with respect to each other (Figure 5.4). Then we plot an equilateral triangle with a side equal to unity along these axes so that the points of origin are located in the point of intersection of the heights, and obtain points A_1; A_2; A_3. Connecting them, we obtain an equilateral triangle $A_1A_2A_3$. The point A_1 is obtained for $P_1 = 1$ and $P_2 = P_3 = 0$. The line A_1A_2 corresponds to a binary mixture when $P_1 + P_2 = 1$ and $P_3 = 0$.

If we lay off the content of one component of a three-component mixture along the P_1 direction, of the second component along the P_2 direction and of the third component along the P_3 direction, we obtain a certain point M located inside the triangle $A_1A_2A_3$. Conversely, any point located inside the triangle can always be projected on three axes.

It can be shown that the sum of axes segments for all internal points of the triangle always equals unity. For that purpose, we plot three lines parallel to three sides of the triangle from point M and obtain three segments a_1; a_2; a_3. We obtain an equilateral triangle BA_2C, the side A_2 of which equals $A_2c = a_1 + a_2$, and if we add a_3 to this value, we obtain $A_2A_3 = a_1 + a_2 + a_3$. By definition, $A_2A_2 = 1$.

It can be easily shown that, in this case too, the composition uncertainty is maximal when all three components are equiprobable, i.e.

$$P_1 = P_2 = P_3 = \frac{1}{3}$$

The maximum uncertainty value for this case is written according to Eq. (5.11)

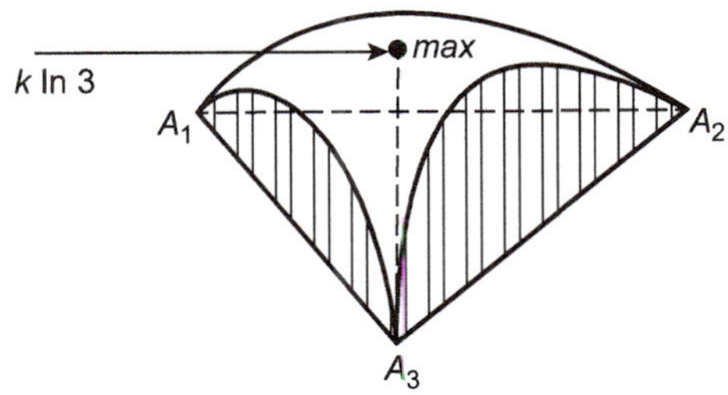

FIGURE 5.5 Entropy surface for a three-component mixture.

$$H_s(x) = -\left(\frac{1}{3}\log\frac{1}{3} + \frac{1}{3}\log\frac{1}{3} + \frac{1}{3}\log\frac{1}{3}\right) = -\log\frac{1}{3} = \log 3$$

To reduce to unity the estimation of the composition heterogeneity for a three-component mixture, it is necessary to take the number 3 as the logarithm base. Then

$$H_s(x)_{\max} = \log_3 3 = 1$$

An axonometric view of the three-dimensional shape obtained in this case is shown in Figure 5.5. We obtain a curve analogous to that of the two-component mixture along each side of the triangle (Figure 5.3). The maximum is reached in the apex of a cupola above the triangle centre.

For any multicomponent mixture, the uncertainty acquires the maximum possible value also at equal contents of all components of the mixture. For instance, for the matrix composition shown above it equals

$$H_s(x)_{\max} = -\sum_{i=1}^{n}\sum_{j=1}^{m}\frac{1}{mn} = -\log\frac{1}{mn} = -\log\frac{1}{mn}\sum_{i=1}^{n}\sum_{j=1}^{m}\frac{1}{mn} = -\log\frac{1}{mn} = \log mn = \log_{mn} mn = 1$$

Thus, entropy maximum for a mixture of any composition can be reduced to unity. It is important for the analysis of mass distribution in any mixture and for the evaluation of separation effect.

CHAPTER

Two-Phase Flow as a Statistical System

6

CHAPTER OUTLINE

1. Substantiation of physical analogy 91
2. Model of particle collisions in a two-phase flow 94
3. External interaction model in a two-phase flow 98
4. Statistical model substantiation 101

1. Substantiation of physical analogy

Gaseous systems can be considered as a distant analogue of two-phase flows. It is known that for one physical phenomenon to be an analogue or model of another, the coincidence of all aspects of their behaviour and characteristics is not obligatory. To assume that two seemingly different physical phenomena are analogous, it is sufficient to reveal at least one common characteristic.

A classic example is a pendulum and an electrical oscillatory circuit. Physically, in fact, there is nothing in common between them. However, their corresponding characteristics are analogous. At the pendulum swing, its deflection from the vertical axis is of sinusoidal character, just as electric current variation in an oscillatory circuit. Long ago, it provided a basis for a successful analysis of one phenomenon by analogy with another.

It can be soundly asserted that the similarity between two-phase flows and kinetic theory of gases is at least the same as between a pendulum and an electrical oscillatory circuit.

As a rule, an immense quantity of particles takes part in two-phase flows even at insignificant concentrations. By way of example confirming this fact, the following figures can be given. At the separation in air flow with the solid phase concentration $\mu = 2.2\ \text{kg/m}^3$, at the average particle size $x = 30\ \mu\text{m}$, one cubic metre of air contains $N = 5.7 \cdot 10^{16}$ particles, and at the average particle size $x = 100\ \mu\text{m}$, $N = 2 \cdot 10^{15}$ particles. Taking into account the fact that up to 40,000 ÷ 50,000 cubic metres of air pass through various apparatuses with two-phase flows of average productivity, it is clear that a sufficiently large number of particles is involved in such processes, so that they can be considered as a statistical system. One also has to take into account that solid particles occupy a relatively small volume in the flow. In both examples, the relative magnitude of this volume is only $\beta = 0.00008$. This points to the fact that despite a great number of particles, the distance between them is significant.

The kinetic theory of ideal gases is based on the hypothesis of molecules representing ideal solid spheres moving at large distances from each other. Short periods of time during which the molecules

Entropic Invariants of Two-Phase Flows. http://dx.doi.org/10.1016/B978-0-12-801458-5.00006-X

interact are considered collisions. Particles in a two-phase flow are also rather far away from each other. It is experimentally established that solid particles in a flow interact intensely enough. According to these data, one square centimetre of a reference particle surface experiences up to 700–800 collisions per second.

The only hope for a deeper insight into the mechanism of two-phase flow is connected to the fact that its most general laws can be realised only using statistical methods similar to the kinetic theory of gases. All other approaches used until recently were absolutely useless because of the complicated character of two-phase flows accompanied by a broad range of random factors.

As known, a basic feature of the statistical approach is that it attempts to determine the state of an entire system on the whole, and not of isolated objects constituting this system.

The statistical model of two-phase flows in separation regimes is aimed at solving the following basic problems that yet remain unsolved:

1. In modern theories based on so-called "velocity hypothesis", the problem of connection between the main factors of the process (medium velocity, solid phase composition, particle size and density, apparatus design) and the distribution of different size classes in separation products is not even considered. A revealed functional connection between them should form the basis of any separation theory, even an approximate one. Only such a theory can help us substantiate the final objective of the prediction, namely, obtaining compositions of separation products by computational method.
2. It is problematic to determine separation results, which depend on the influence of solid phase granulometry and consumed concentration, on the basis of existing theories.
3. The majority of attention in the theory of separation processes is usually focused on the flow effect on solid particles. However, solid phase also affects a moving medium. This effect is rather significant, and it also should be considered when developing the process theory.
4. A very important aspect of correctly developing the theory is to determine optimal flow velocity. It is usually assumed that it equals the final settling velocity of particles of boundary size. It is implied that separation takes place in the suspended state, assuming that the velocities of hovering and settling particles are identical. This is, however, far from being true. First, these parametres are of different physical natures – hovering velocity relates to the flow motion, while settling velocity relates to the solid particle. Second, drag resistance of a particle settling in a motionless medium differs from that of a hovering particle by a value affected by the medium flow regime. Third, a certain structure (velocity profile) is always established in a flowing medium. Local velocity values in a transverse channel flow vary and can significantly differ from mean ones chosen for estimations.
5. Within the generally accepted "velocity hypothesis", there is no basic difference between separation processes realised in liquid and gaseous media. Both are usually described by similar empirical relations. However, the difference between them is rather basic due to a higher density of liquid and a higher mobility of gas. Therefore, there are good reasons to assume that main regularities for wet and dry separation are basically different, although they can coincide to some extent.
6. A very important issue is that the "velocity hypothesis" requires such organisation of the separation process that the mean flow velocity equals the final settling velocity (or hovering

velocity). According to the operation experience and to the study of vertical cascade classifiers, the optimal flow velocity value is not only a function of particle size and boundary size, but also depends on the place of material feed into the apparatus and on the apparatus length.

Principal tools of statistical analysis are based, as shown in Chapter 1, on various aspects of the probability theory. This theory is one of the branches of mathematical analysis and, in a narrower sense, of the measure theory. In its most rudimentary form, it is connected to combinatorial calculus and other fields of mathematics important for our analysis.

The potential of probabilistic analysis can be illustrated by another elementary example. Consider a mined ore mixture consisting of particles of broad size and density ranges. If the quantity of ore reaches hundreds or thousands of tons, at first sight, such a system is totally chaotic. However, if we apply a probabilistic approach to this system, an elementary analysis can provide clear and well-founded granulometric characteristics or a complete description of the composition using a matrix.

This example does not establish the potential of a probabilistic approach to the analysis of phenomena and processes, given that it is hardly possible to define the applicability boundaries of such an approach.

Thermodynamics considers gases of neutral molecules under a pressure up to hundreds of atmospheres as practically ideal. Even at such pressures, the probability of triple collisions with the participation of three molecules is exceedingly small in comparison with the probability of pairwise collisions.

The probability of triple collisions in two-phase flows has proved to be insignificant, too. Besides, here the probability of collisions of similar particles is also low, because mean velocity of similar particles in through flows is approximately the same. Their interaction does not lead to noticeable effects of momentum and energy redistribution. In this respect, collisions between particles of different sizes, which can be oppositely directed in critical regimes, lead to the greatest effect.

The kinetic theory of gases is based on the description of molecular motion using classic Newton's mechanics. On the whole, the kinetic theory is a theory of collisions and mean free path of particles between two successive collisions. Note that main results of the kinetic theory of gases have been confirmed by all experimental facts obtained during the nearly 150 years since its creation. This theory has been successfully applied not only in its own field, but also in many other fields of science even distantly related to the kinetic theory of gases.

Therefore, it seems of interest to analyse the character of two-phase flows from a similar standpoint. The obtained results of such analysis should be compared with experimental data. The extent of their adequacy will lead to a conclusion about the validity of a statistical approach to two-phase flows.

However, in such an analysis one has to overcome many difficulties. First, dust particles are surely much bigger than molecules. Therefore, differential calculus is inapplicable for modelling the behaviour of such particles and, under certain limitations, only finite-difference parametres can be used. Strictly speaking, because molecule sizes are finite, it is problematic to apply differential calculus to molecules, too.

Second, the shapes of dust particles differ from ideal spheres. Gas molecule shape is not ideally regular either. However, they were considered to be spheres and, on the whole, excellent modelling results were attained. Taking into account the fact that this simplification is incomparably coarser, we assume at first that solid dust particles are regular spheres of various sizes equivalent to particle volumes. In the model under study we are interested only in the projection of particle velocity onto the vertical axis, given that mass distribution is directed vertically upwards and downwards. As further analysis shows, even such a simplified or even primitive model can describe the process results in detail, which is confirmed by an immense number of empirical laws obtained long ago, which remain unexplained until now.

Third, gaseous systems are considered according to statistical theory to be comprised in a closed volume, with equiprobable directions of molecular motion. As for two-phase flows, closed volumes are out of the question, and besides, the resulting flow has a preferential direction.

A determining parametre of the statistical approach in the theory of gaseous systems is the potential energy of the particles' continuum depending on the temperature. It is clear that a two-phase flow is provided with energy by other sources. As for temperature, its change does not practically affect main parametres of such flows.

One more observation: Here we have to reject generally accepted parametres of a system (temperature, energy, heat capacity, work, enthalpy, etc.) and to introduce new ones defining potential extraction, probability of particles motion direction, flow mobility, chaotizing factor, etc. (See two books by the author published in 2006 and 2010.)

At the same time, it is noteworthy that ideas and methods of gas mechanics are used here to examine the two-phase flow model. In this study, certain dependencies are obtained, which structurally resemble thermodynamic laws. Although they, as well as the respective parametres, have an absolutely different physical sense, we give them names accepted in statistical mechanics (e.g., Boltzmann factor, Gibbs factor, entropy, statistical sum).

2. Model of particle collisions in a two-phase flow

Due to many reasons, solid particles in a real two-phase flow can deflect from rectilinear trajectories, changing their instantaneous velocity. Therefore, it is necessary to average some details of the interaction in order to retain important information on the probability of the fact that two particles with the velocities v_i and v_j in the beginning of their interaction scatter after their interaction with the velocities v_i' and v_j', respectively. Such an assumption is perfectly relevant, because in stochastic problems, operations with values characterised by a probabilistic distribution are usually replaced by operations with their mathematical expectations.

Clearly, in such a situation initial and final velocities of particles have a markedly probabilistic character of distribution within a certain range of values.

There are two types of impact interactions in a two-phase flow. All collisions between particles, which are called internal collisions in the system, belong to the first type. Solid particle interactions with channel walls enclosing the flow or with internal devices inside the channel (if any) are called external interactions and belong to the second type.

During the collisions, mass, momentum and energy conservation laws hold. Because particles in a flow do not undergo any transformations, the mass of two particles before and after a collision remains unchanged.

To examine interaction details, we single out a segment of Δl size in a vertical ascending flow. Denote the equivalent cross-section diametre of the channel by D_e (Figure 6.1). As noted, because it is impossible to operate with differentials, we operate with differences between various values. Therefore, the segment Δl should be large enough to hold a representative number of solid particles and small enough in comparison with the scales of velocity and concentration variations of these particles.

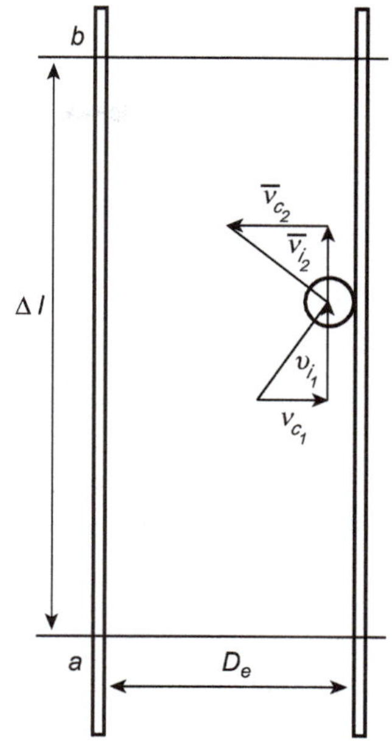

FIGURE 6.1 Velocity transformation at a particle collision with the wall.

We examine the interaction of two fractions – fine and coarse particles – and introduce the following notations: r_i;m_i;v_i are the radius, mass and mean axial velocity component of a fine particle, and r_j;m_j; v_j– the radius, mass and analogous velocity of a coarse particle.

In a unit time, a coarse particle can collide with those fine particles whose centres are located at that moment within a cylinder with the base equal to the collision area and the height equal to the difference between particle velocities $(v_i - v_j) = h$. The collision area is

$$S = (r_i + r_j)^2 \cdot \pi \tag{6.1}$$

Denote particle velocities after the collision v_i' and v_j'. Momentum conservation for the internal interaction is expressed by the relation

$$m_i v_i + m_j v_j = m_i v_i' + m_j v_j' \tag{6.2}$$

According to the theorem of the gravity centre velocity of a system, one can assert that in this case it remains unchanged, that is,

$$v_0 = \frac{m_i v_i + m_j v_j}{m_i + m_j} = \frac{m_i v_i' + m_j v_j'}{m_i + m_j} \tag{6.3}$$

In this case, the total kinetic energy of a system of two particles remains unchanged, that is,

$$\frac{m_i v_i^2}{2} + \frac{m_j v_j^2}{2} = \frac{m_i v_i'^2 + m_j v_j'^2}{2} \tag{6.4}$$

Particle velocities after the collision can be unambiguously determined from expressions Eqs (6.3) and (6.4):

$$\left.\begin{aligned} v_i' &= v_j + \frac{m_i - m_j}{m_i + m_j}\left(v_i - v_j\right) \\ v_j' &= v' + \frac{m_i - m_j}{m_i + m_j}\left(v_i - v_j\right) \end{aligned}\right\} \tag{6.5}$$

In a case of absolutely elastic bodies, particle velocities after a collision change their sign only, because, as follows from Eq. (6.5),

$$v_i' - v_j' = v_j - v_i \tag{6.6}$$

A total shock pulse affecting a cluster of fine particles per unit time related to a unit flow volume can be determined as

$$Q_f = \Delta N_f \Delta v_i \tag{6.7}$$

where ΔN_f is a mass of fine particles colliding with coarse ones; Δv_i– fine particles velocity change after the collision. According to Eq. (6.5), this value can be defined as

$$\Delta v_i = v_i' - v_i = \frac{2m_j}{m_i + m_j}\left(v_j - v_i\right) \tag{6.8}$$

To determine ΔN_f, we take into account the following. To determine the probability of particle collisions $P(x)$, one can assume that coarse particles are motionless, and fine ones move at certain relative velocities. The probability $P(x)$ can be defined as a ratio of all collision areas in the flow cross-section to the value of said cross-section, that is,

$$P(x) = \frac{4\sum S}{\pi D_e^2} \tag{6.9}$$

where $\sum S = \Delta n_j (r_i + r_j)^2 \pi$; n_j is an average number of coarse particles in the flow cross-section

$$n_j = \frac{G_j \cdot 2r_j}{m_j g \Delta l} \tag{6.10}$$

where G_j is the weight of coarse particles in the segment Δl. Only those fine particles that are not farther than the distance h from this layer collide with these particles during a unit time, that is,

$$N_i = \frac{G_i\left(v_i - v_j\right)}{m_i g \Delta l} \tag{6.11}$$

where G_i is the weight of fine particles in the segment Δl. Not all fine particles take part in a collision, but only a part of them determined by the collision probability within the cross-section.

$$\Delta N_i = N \cdot P(x) = \frac{4\left(r_i + r_j\right)^2 \pi G_i n_j N\left(v_i - v_j\right) 2r_j}{D_e^2} \tag{6.12}$$

The total number of collisions in the entire volume of the flow under study is

$$\Delta N = \frac{\Delta N_i \Delta e}{2r_j} = \frac{4(r_i + r_j)^2 (v_i - v_j) n_j N_i \Delta l}{D_e^2} \tag{6.13}$$

The values n_j and N_i correspond to the number of particles of both fractions in a unit height of the flow under study. Passing from the number of particles to the consumed concentration of fine and coarse particles, Eq. (6.13) can be transformed into

$$N' = \frac{\pi(r_i + r_j)^2}{m_i m_j} (v_i - v_j) \mu_i \mu_j \tag{6.14}$$

The relation (6.14) shows the total number of collisions in a unit volume. Taking into account Eqs (6.8) and (6.14), the relation (6.7) can be written as

$$Q_f = \frac{\pi(r_i + r_j)^2}{(m_i + m_j)} (v_i - v_j)^2 \mu_i \mu_j \frac{m_i m_j}{(m_i + m_j)} \tag{6.15}$$

If we single out all constants in Eq. (6.15), then

$$Q_f = c\mu_i \mu_j (v_i - v_j)^2 \tag{6.16}$$

where $c = \frac{\pi(r_i+r_j)^2 m_i m_j}{(m_i+m_j)^2}$

We analyse the relation (6.16). At a steady motion, mathematical expectation of fine and coarse particles is

$$\left. \begin{aligned} v_i &= w - w_{i0} \\ v_j &= w - w_{j0} \end{aligned} \right\} \tag{6.17}$$

where w is flow velocity; w_{i0}, w_{j0}– hovering velocities of fine and coarse particles. Hence, for specified particles

$$v_i - v_j = w_{j0} - w_{i0} = const$$

It means that the total force of the internal particles interaction in a flow is

$$Q = c_1 \mu_i \mu_j \tag{6.18}$$

This result is somewhat paradoxical. The total shock pulse in a flow is independent of the flow velocity and is determined only by consumed concentrations of colliding particles.

Equation (6.12) show how many fine particles interact with coarse ones per unit time. This value amounts to

$$z_j = \frac{\Delta N_i}{n_j} = \frac{\pi(r_j + r_i)^2}{m_i} (v_i - v_j) \mu_i \tag{6.19}$$

During the time interval Δt a particle passes along a zigzag path $v_j \Delta t$ long. The free path length of a particle can be defined as a ratio of the particle path to the number of collisions that have occurred over this path

$$\lambda_j = \frac{v_j \Delta t}{z \Delta t} = \frac{v_j}{z} = \frac{m_i v_j (w - v_{j0})}{\pi (r_i + r_j)^2 (v_i - v_j) \mu_i} \tag{6.20}$$

Equation (6.20) shows that the free path length of a particle of a certain size class in a two-phase flow is proportional to the flow velocity and inversely proportional to the concentration of particles of another class in said flow.

3. External interaction model in a two-phase flow

It is experimentally established that the motion of particles in a two-phase flow is not parallel to its axis. The existence of random disturbing factors in such flows causes a radial component in their velocities. As demonstrated in many researches, the trajectory of particles with the size above 1 mm slightly depends on turbulent pulsations of the medium. The main reason for particles to deflect from rectilinear paths lies in their mass impact interactions.

Among recent works on the dynamics of two-phase systems, only some take into account solid phase interaction with channel walls. In most cases, such research results in empirical relations only.

The velocity of particles of any size in a polyfractional two-phase flow can be represented by two components (Figure 6.1). The radial component causes solid phase interaction with vertical channel walls. At each collision with the wall, a part of the kinetic energy of a particle is lost, and this loss is compensated afterwards by the carrier flow energy. Hence, it is clear that solid phase interaction with channel walls leads to an increase in the total drag resistance of the flow.

At a stable granulometric composition of the dispersed component, its interaction with the walls grows with increasing flow velocity and solid phase concentration up to the value $\mu = 5 \div 6\ kg/m^3$. At a significant concentration growth, radial displacement of particles is reduced due to their suppression because of a higher flow constraint.

Hence, one can conclude that another force acts against each narrow class motion, increasing the total flow resistance. We determine its value from the accepted model concepts. For a segment with the size Δl in Figure 6.1, one can determine the weight of solid particles of $i-$ th size class as follows:

$$G_i = \rho_i g F_0 \cdot \Delta l = \rho_i g V \tag{6.21}$$

where ρ_i is mass concentration of the $i-$ th class in the flow; $g-$ gravity acceleration; $V-$ segment volume. The force acting on the wall can be determined on the basis of the second Newtonian law. It is equal and of the opposite sign to the change in the momentum of particles colliding with the wall per unit time. No matter how the particle velocity is directed, it can be always split into components, one of them being perpendicular to the straight channel wall, and another parallel to the axis flow. In Figure 6.1, the axial component of the particle velocity before the collision is denoted by v_{i1}, and radial component by v_{r1}.

The momentum of such a particle in radial direction is $m_i v_{ri}$. After the collision, the particle acquires a radial component v_{r2}. The value of this velocity is determined by elastic properties of the particle and the wall, and also by the surface state at the point of contact. Denoting a generalised elasticity coefficient of a particle (taking into account the influence of the wall) by K, we can write

$$v_{r2} = -v_{r1} \cdot K$$

The change in the momentum at a collision amounts to

$$\Delta Q = m_i v_{r1} - m_i v_{r2} = m_i v_{r1}(1 + K)$$

The weight of particles of a narrow size class under study that reach a channel wall in a unit of time is proportional to the material content in a unit volume of the channel, to the mean radial component of their velocity and to the contact area value; that is,

$$\Delta G_i = \psi g \rho_i v_{r1} S \Delta l \quad (6.22)$$

where ψ is proportionality factor; $S-$ channel cross-section perimetre.

If n particles of the size class under study reach the wall per unit time, it leads to a change in the particles momentum:

$$Q = \sum_n \Delta Q = \sum m_i v_{r1}(1 + K) = n m_i v_{r1}(1 + K)$$

Only the particles located not farther than v_{r1} from the wall can reach it per unit time, that is, all those contained within the volume of an annular cylinder with an external generatrix $S \cdot \Delta l$ and ring base width equal to v_{r1}. The average number of particles per unit space is

$$n_0 = \frac{\rho_i}{m_i} = \frac{G_i}{F_0 \Delta l m_i g} \quad (6.23)$$

The volume under study comprises

$$N_0 = n_0 v_{r1} S \Delta l = \frac{G_i v_{r1} S}{F_0 m_i g} \quad (6.24)$$

particles. We can assume that due to the random character of the process under study, half of the particles in the specified space move towards the wall, and another half away from the wall. However, because of the absence of direct experimental data confirming this assumption, we can write in the general case

$$N_1 = \varphi N_0$$

where $\varphi < 1$ (according to the statistical meaning, $\varphi = \frac{1}{2}$).

Mean momentum acting of the surface per unit time is

$$Q_i = \varphi \frac{G_i v_{ri} S}{F_0 m_i g} m_i v_{r1}(1 + K) \quad (6.25)$$

The value of this impact per unit surface amounts to

$$Q_{i0} = \frac{Q}{S \cdot \Delta l} = \varphi \frac{G_i v_{r1}^2 (1 + K)}{g F_0 \cdot \Delta l} = \varphi \frac{G_i v_{r1}^2}{g V}(1 + K) \quad (6.26)$$

According to the physical meaning, it can be assumed that

$$\varphi = \frac{1}{2}; \quad K = 1$$

Then

$$Q_{i0} = \frac{G_i}{V} v_r^2 = \mu_i v_r^2 \tag{6.27}$$

Obviously, the total impact amounts to

$$Q_0 = \sum_i Q_{i0} = \sum_i \mu_i \cdot v_{r1}^2 = \mu \cdot v_{r1}^2 \tag{6.28}$$

where μ is the total solid phase concentration in the flow.

The normal impact to the channel wall causes a friction force

$$\tau = fQ_0$$

where f is the friction factor. It follows from numerous studies that the radial component of particle velocities is mainly determined by the particle velocities, that is

$$v_{r1} = \psi \cdot v_i \tag{6.29}$$

Then, taking Eq. (6.26) into account, one can write

$$\tau_i = \varphi f \psi \frac{G_i v_i^2 (1+K)}{g \Delta F l}$$

Combining all constant factors in this relation,

$$\lambda_i = \frac{1}{2} \varphi f \psi (1+K),$$

one can write

$$\tau_i = \lambda_i \frac{G_i v_i^2}{2gF_0 \Delta l} \tag{6.30}$$

This dependence was obtained purely experimentally. It is used for calculating drag in pipelines used for pneumotransport.

Friction resistance due to all size classes is

$$T = \sum \tau_i = \sum \frac{\lambda_i G}{2g} \cdot \frac{v_i^2 S \Delta l}{F \cdot \Delta l} = \frac{\lambda_r}{2gD_e} \sum G_i v_i^2 \tag{6.31}$$

Friction coefficients in the form of the relation

$$\lambda = \lambda_i \frac{v_i}{w}$$

were experimentally determined for pneumotransport conditions by Gasterschtadt, Zegler, Urban, Papai and many others.

This points to the fact that approaching the physical phenomenon under study from the standpoint of the kinetic theory is quite admissible. It leads to results obtained in the study of two-phase flows, which were checked and confirmed more than once. We have not obtained any new generalising invariants, but only proved the validity of previously obtained empirical relations. At the same time, this confirms the permissibility of the approach to such a flow from a statistical standpoint used in the kinetic theory of gases.

4. Statistical model substantiation

An enormous number of solid particles take part in the flow under study. As shown in the model of the process, to the first approximation, they can be considered as solid spheres, by analogy with molecular presentation in the statistical theory of ideal gases. This gives grounds to apply some approaches of the kinetic theory to the process under study. Although two-phase flows have specific peculiarities, nevertheless, mass character of the motion and interactions of particles with each other and with channel walls give hopes to construct a successful model. The object of this model should be a connection between the general behaviour of the continuum and of separate particles forming the continuum. The target of our research is to analyse such connections.

Consider mass motion of solid particles in an ascending medium flow. Here, flow regimes that provide conditions for counter-directed motion of particles of polyfractional solid phase are of greatest interest. Empirical results of mass distribution in such a flow were presented earlier. Boltzmann's theory of gases shows that in order to develop a successful model of a process, it is not obligatory to describe the process thoroughly. It is sufficient to establish an elementary scheme of the process, but the scheme should reflect its essence.

First, examine the behaviour of one narrow size class in the flow. It has been empirically established that in a certain concentration range, each narrow class of particles behaves autonomously in a flow, irrespective of the behaviour of other classes. Here we assume that this class consists of identical particles.

For constructing the model, consider a certain amount of such particles moving in both directions, up and down, together with the ascending flow through a certain volume of the channel. This volume or its part, together with solid particles limited at different heights by two parallel planes, can be considered as a statistical system. In such a system, we are interested in mass distribution of the solid phase in both directions. We neglect other characteristics of the process, such as particles' interaction with one another and with channel walls, local inhomogeneities of the concentration fields, medium motion regime, and other factors characterising a two-phase flow. We also neglect actual velocities of particles and their direction in space and examine only the projections of the velocities of these particles to the vertical axis. We are not interested in the values of these projections – our attention is mainly attracted by their direction.

A system of the directions of velocity projections of all the particles in the volume under study forms a statistical system. We examine this system in a stationary state only, rather than in general. It means that the location of a particle in any volume element is time independent, which predetermines time independence of all physical properties of the system. In hydrodynamics, such flow is considered as stationary.

Determine the separation factor having a different orientation in vertical direction. This parametre can be written as

$$z = mgd\sum_{N}(\pm 1) + (\pm 1) + (\pm 1) + \ldots + (\pm 1)$$

Using the ergodic principle, this situation can be examined in a somewhat different aspect. A similar result can be obtained by observing one fixed particle performing N successive independent steps up and down along the vertical axis. As a result, this particle is displaced by N steps in one of the directions. Such a situation is called random walk.

We assume that the size class under study representing a stationary statistical system consists of N equivalent Uparticles. Each of these particles has a chance or probability to go up equal to ρ, or to go down equal to $-g$. Clearly, for a single particle,

$$\rho + g = 1$$

According to the invariance of separation curves and the ergodic principle, one can assert that n particles out of N have a probability to go up and $(N-n)$ to go down. Here one deals with a discrete distribution of probabilities, because n can assume any integer value from 0 to N.

While examining characteristics of a system, it becomes necessary to use parametres for upward n and downward $(N-n)$ orientation, which entails certain inconveniencies. It turns out that it is possible to reduce the number of such parametres by introducing another index for potential piece-wise accounting for the direction of particles in connection with their quantity as follows:

$$\left.\begin{aligned} \varepsilon_f &= \frac{N}{2} + z \\ \varepsilon_c &= \frac{N}{2} - z \end{aligned}\right\} \tag{6.32}$$

where z is the separation factor. Certainly, at an equiprobable separation with $\varepsilon_f = \varepsilon_c = 0.5$, the magnitude of $z = 0$.

In the physical sense, the separation factor z equals the number of particles corresponding to the deviation of ε_f and ε_c values at a departure from the equiprobable distribution. These parametres differ by a constant only, and their derivatives are the same modulo. Because every particle has two orientations, the number of system states is 2^N. Here the possible value of potential extraction is only $(N+1)$. For instance, for two particles we can obtain three values of separation factor:

1) $\uparrow\uparrow$ – both particles are oriented upwards $(z = +2)$;
2) $\downarrow\downarrow$ – both particles are oriented downwards $(z = -2)$;
3) $\uparrow\downarrow$ and $\downarrow\uparrow$ – the particles possess opposite orientations $(z = 0)$. Note that the latter values of the system are self-similar.

Thus, the number of states exceeds the number of possible values of potential extraction. For example, for $N = 10$, there are $2^N = 1024$ states of the system, which corresponds to only 11 different values of potential extraction.

It is convenient to assume that N is an even number. We study cases where N is very large, and therefore, it is inessential whether N is even or odd. Note that the piece-wise difference in the orientation of particles is always even and amounts to

$$\left(\frac{N}{2} + z\right) - \left(\frac{N}{2} - z\right) = 2z$$

This parametre varies within the range

$$-\frac{N}{2} \leq z \leq +\frac{N}{2}$$

The probability for all of N particles to become oriented upwards equals ρ^N. The probability for one particle to become oriented downwards, while all others are oriented upwards, equals $\rho^{N-1}g$. The number of ways of distributing N particles equals $N!$ These particles are distributed into two groups, in which n particles are oriented upwards and $(N-n)$ downwards. The total number of distribution ways or complexes can be obtained by dividing $N!$ by the product $n!(N-n)!$:

$$\varphi = \frac{N!}{n!(N-n)!}$$

Multiplying this by $\rho^n g^{N-n}$, we obtain the probability of any distribution between two directions. We obtain the desired distribution in the form

$$\varphi = \frac{N!}{n!(N-n)!}\rho^n g^{N-n} \tag{6.33}$$

The sum of all probabilities covers all the possibilities:

$$\sum_{n=0}^{N}\varphi = \sum_{n-0}^{N}\frac{N!}{n!(N-n)!}\rho^n g^{N-n} = (\rho + g)^N \tag{6.34}$$

Mathematical expectation of the quantity n is

$$\langle n \rangle = N_\rho \tag{6.35}$$

The variance of this quantity is

$$\sigma_n^2 = N\rho g \tag{6.36}$$

A square root of this quantity is called standard deviation

$$\sigma_n = \sqrt{N\rho g} \tag{6.37}$$

Clearly, the standard deviation is maximum for $\rho = g = 0.5$. Relative deviation

$$\frac{\sigma_n}{N} = \sqrt{\frac{\rho g}{N}} \tag{6.38}$$

is a measure of $\frac{n}{N}$ deviation from the expected probability ρ in any sequence of N particles. The dependence of Eq. (6.33) type is a binomial distribution, and therefore, we can use a binomial expansion for the dependence Eq. (6.34):

$$(\rho + g)^N = \rho^N + N\rho^{N-1} + \frac{1}{2}N(N-1)\rho^{N-2}g^2 + \cdots + g^N$$

This expression can be more compactly presented as the dependence Eq. (6.33). In this dependence, it is more convenient to search for states within different limits. The quantities n and $(N-n)$ represent, as if piece-wise, particles extraction into the upwards and downwards directions. As demonstrated by Eq. (6.32), unambiguous distribution in both directions requires limits from $-\frac{N}{2}$ to $+\frac{N}{2}$. Taking this into account, Eq. (6.34) can be represented by an identical dependence:

$$\sum_{-\frac{N}{2}}^{+\frac{N}{2}} \varphi = \sum_{-\frac{N}{2}}^{+\frac{N}{2}} \frac{N!}{\left(\frac{1}{2}N + z\right)\left(\frac{1}{2}N - z\right)} \rho^{\frac{N}{2}+z} g^{\frac{N}{2}-z} \tag{6.39}$$

Taking this into consideration, we pass to further calculations, which require additional clarifications.

CHAPTER

Main Statistical Parameters of a Two-phase Flow

7

CHAPTER OUTLINE

1. Substantiation of a statistical system for solid particles continuum 105
2. Generalising parameter for a two-phase flow 106
3. Generalising invariant for two-phase flows 113
4. Definition of chaotising factor for two-phase flows 117

1. Substantiation of a statistical system for solid particles continuum

Methods of classical mechanics are, in principle, applicable to an array of solid particles moving in a flow. Here the behaviour of the entire continuum can be specified by determining the behaviour of each separate particle, for which one can write, according to the second Newton law,

$$\frac{dx_i}{dt} = v_i; \quad \frac{dv_i}{dt} = F_i; \quad \text{hence,}\ \frac{d^2x_i}{dt^2} = F_i$$

where F_i is the force acting on the i-th particle, x_i is a coordinate of the i-th particle, v_i is the velocity of the i-th particle.

This expression shows all the parameters of the i-th particle along one axis x. To obtain a complete pattern of motion, analogous equations are needed for y and z axes. Theoretically, if we knew coordinates and momentums of all the particles, we would know all about the continuum.

The behaviour of particles of a narrow size class consisting of N particles should be described in Cartesian coordinates by a vector of $6N$-dimensional system. Each of N particles is described by six components of this vector, three coordinates of a particle and three components of its momentum.

In a general case, F_i value is composed of gravitational forces and forces developed by the flow as well as i-th particle interaction with other particles and with walls confining the flow. To solve the obtained system, it is necessary to define $6N$ initial values of the main parameters.

It is perfectly clear that this problem is insoluble not only due to a great number of particles, but also because all these equations are interconnected, as the force acting on each particle at each moment of time is a function of the position of the rest of the particles of the system, i.e.

$$F_i = f(x_j) \quad (j = 1, 2\ldots, N) \quad \text{for} \quad i \neq j$$

The solution of this problem can be sought by statistical methods only. Obviously, here not only initial conditions of particles locations, but also interaction features should be averaged. Practical

Entropic Invariants of Two-Phase Flows. http://dx.doi.org/10.1016/B978-0-12-801458-5.00007-1

results of such averaging are reduced to operations with probabilities instead of certainties. Within such approach, we cannot discuss a definite position and velocity of a specific particle, but only the probabilities of the realisation of its various positions and velocities. One should not expect that such averaging will lead to a process model similar to a real flow. The model will only resemble the original reflecting some of its real characteristics. For their more definite correlation, the following comparison can be suggested. The model can resemble the original as a caricature. The more successful the model, the more features of the original are recognisable in it. This recognisability determines the reliability of the results obtained using the model. A probabilistic approach to such a system forms a basis for hopes to formulate entropy parameter for it. It will lead to a significant progress in physics of two-phase flows, as it is generally accepted that entropy appears as an invariant of transformations in the theory of dynamic systems with probabilistic realisation.

To solve this model, it is necessary to define a certain generalising parameter characteristic of two-phase flows only which can play the part of a response surface at the change in the system parameters in $(6N + 1)$-dimensional space.

2. Generalising parameter for a two-phase flow

To find a generalising parameter for a system consisting of N identical particles moving in a two-phase flow, we restrict ourselves by the following model.

To develop the model, we examine a certain number of identical solid particles moving vertically upwards and downwards together with an ascending flow through a limited volume of space. This volume can be considered as a natural space of the entire separator or its part limited by two horizontal planes at a distance l from one another. In such a flow, we are interested only in the redistribution of the initial material, more exactly, in the fractional extraction of different size classes into upper and lower products. Therefore, we neglect particle velocity value and examine only velocity direction (i.e. its projection onto the vertical axis) for each particle (Figure 7.1). This corresponds to the essence of the process under study and therefore our model is based on the difference between particle velocities directions.

Naturally, it is a very simplified model. However, it is not simpler that the model of moving medium examined previously, which served the basis for all principal regularities and similarity criteria obtained for hydraulic systems.

In this model, particle velocity direction can have only two orientations: upwards or downwards. Note that the orientation probability for each particle is independent of the orientation of other particles. Besides, we are not interested in any other process parameters, such as actual velocity direction, velocity value, particles interaction with each other and the walls confining the flow, local concentration inhomogeneities, except for the instantaneous particle velocity projection onto the vertical axis. Thus, the accepted model of the process reflects the main idea of separation: opposite direction of particles in a flow. Taking into account velocity direction only, we introduce notations for upward (a) and downward (b) direction according to Figure 7.1.

Note that the probability of a zero projection of the velocity onto the vertical axis is vanishingly small, as at that moment the particle must be motionless with respect to the walls or move strictly normally to a wall. The probability of such a state approximately equals the probability of a coin to

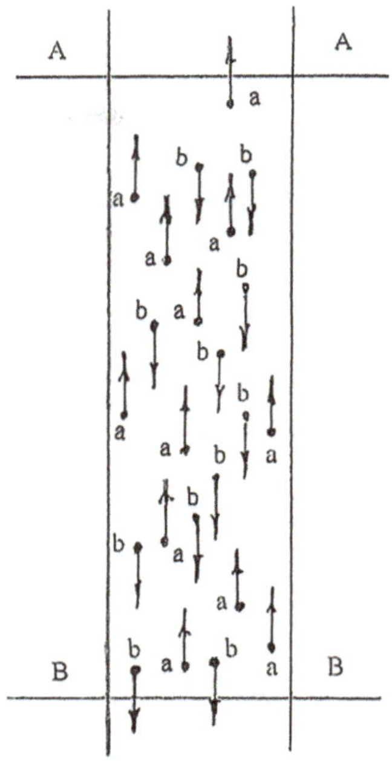

FIGURE 7.1 Statistical model of a two-phase flow.

stay on its edge when thrown. From the standpoint of the process under study, this probability is so small that it is not of interest.

In principle, for other kinds of two-phase flows, the principal axis can be arranged otherwise, e.g. horizontally for horizontal or centrifugal flows or aslant for inclined flows.

It is most natural for gravitational separation to assume that the principal axis is vertical. We call a system the totality of all particles passing in both directions through a flow space limited by height. In Figure 7.1, the system is limited by lines A and B.

The object of our study is only a system with a steady process, and not any system. As there is no permanent material accumulation in the selected space and because the total output of both products in a stationary process always equals the initial feed, we can assume that the number of particles in a selected volume is approximately constant.

The totality of such particles passing through a limited space is considered in the following section as a statistical system. Here it is necessary to use only one notion of statistical mechanics, namely, the notion of a stationary state of a system. In a stationary state of a physical system, the probability of finding a particle in any unit volume is time-independent. This state can be defined more rigorously as the state of a system, in which all its observable physical properties are explicitly time-independent.

In such a state of the system with a constant (in time) velocity of the ascending flow of the medium, fractional separation of various size classes is not explicitly time-dependent, i.e. its fluctuations are insignificant and markedly random. As a rule, such a state comes after the lapse of a certain period since the beginning of the process or after external disturbances (relaxation time). Anyway, we accept the possibility of some fluctuations.

The most important thing is that such stationary states of the system under study are countable, although their number can be great.

From the mathematical standpoint, the disorder in a system is determined by the number of different ways of distributing a certain set of objects. The greater the number of such objects, the higher the probability of their random distribution instead of an ordered state.

We count the number of different ways of distributing particles in the model under study to satisfy certain restrictions imposed on the system. For this purpose, we must, first of all, find parameters allowing us to distinguish the particles.

To represent a single state of a system, one can use either a visual image as in Figure 7.1 or a symbolic expression

$$a_1 b_2 a_3 a_4 b_5 a_6 b_7 \ldots a_i b_j \ldots b_N \tag{7.1}$$

Starting with the simplest calculation, we multiply $(a_1 + b_1)(a_2 + b_2)$ for a system of two particles and obtain four possible states

$$(a_1 + b_1)(a_2 + b_2) = a_1 a_2 + a_1 b_2 + b_1 a_2 + b_1 b_2$$

Note that the second and third states of the system are self-similar, as in these states one particle is oriented upwards and another downwards. In the same way, we can multiply the expressions $(\uparrow_1 + \downarrow_!)$ $(\uparrow_2 + \downarrow_2)$. A determining function for three particles gives

$$\begin{aligned}(a_1 + b_1)(a_2 + b_2)(a_3 + b_3) = {} & a_1 a_2 a_3 + a_1 a_2 b_3 + a_1 b_2 b_3 + a_1 b_2 a_3 \\ & + b_1 a_2 b_3 + b_1 b_2 b_3 + b_1 a_2 a_3 + b_1 b_2 a_3\end{aligned}$$

Here, eight states of the system are obtained which corresponds to

$$2^3 = 8$$

and for eight states of the system we obtain only four potential extraction values. The product of N multipliers in Eq. (7.1) can be written without taking into account the numeration of particles, which is not of basic importance from the standpoint of the process under study.

Because in our case, particle velocity projection has only two orientations, the total number of states of a system of N particles is

$$\varphi = (a + b)^N \tag{7.2}$$

In a general case, this dependence can be expanded using binomial theorem

$$(a + b)^N = a^N + N a^{N-1} b + \frac{1}{2} N(N - 1) a^{N-2} b^2 + \ldots + b^N$$

This expression can be written in a more compact way as

$$(a + b)^N = \sum_{k=0}^{N} \frac{N!}{(N - k)!k!} a^{N-k} b^k$$

where k is the summand number. In the range of separation factor variation from $-\frac{N}{2}$ to $+\frac{N}{2}$, this relation acquires the form

$$(a+b)^N = \sum_{-\frac{N}{2}}^{+\frac{N}{2}} \frac{N!}{\left(\frac{1}{2}N+z\right)!\left(\frac{1}{2}N-z\right)!} a^{\frac{N}{2}+z} a^{\frac{N}{2}-z}$$

The structure of this relation reminds that of Eq. (6.39).

The product $a^{\frac{N}{2}+z} b^{\frac{N}{2}-z}$ enumerates all possible separation factors in $-\frac{N}{2} \leq z \leq +\frac{N}{2}$ range, whereas the binomial coefficients show the number of self-similar states of the system with a fixed number of particles oriented upwards or downwards.

We perform calculations under the conditions of $N \gg 1$ and $z \leq \left|\frac{N}{2}\right|$. Determine binomial coefficients

$$\varphi(N;z) = \frac{N!}{\left(\frac{1}{2}N+z\right)!\left(\frac{1}{2}N-z\right)!} \tag{7.3}$$

In this expression, z is any integer between $-\frac{N}{2}$ and $\frac{N}{2}$. Taking a logarithm of the left- and right-hand sides, we obtain the following in Eq. (7.3):

$$\ln \varphi = \ln(N!) - \ln\left[\left(\frac{N}{2}+z\right)!\right] - \ln\left[\left(\frac{N}{2}-z\right)!\right]$$

We examine separate parts of this expression:

$$\ln\left[\left(\frac{N}{2}+z\right)!\right] = \ln\left(\frac{N}{2}!\right) + \sum_{k=1}^{z} \ln\left(\frac{N}{2}+k\right)$$

$$\ln\left[\left(\frac{N}{2}-z\right)!\right] = \ln\left(\frac{N}{2}!\right) - \sum_{k=1}^{z} \ln\left(\frac{N}{2}-k+1\right)$$

and add them up:

$$\ln\left[\left(\frac{N}{2}+z\right)!\right] + \ln\left[\left(\frac{N}{2}-7\right)!\right] = 2\ln\left(\frac{N}{2}!\right) + \sum_{k=1}^{z} \ln\frac{1+\frac{2k}{N}}{1-\frac{2k}{N}} \tag{7.4}$$

Proceeding from the assumption that $\frac{N}{2}-k+1$ approximately equals $\frac{N}{2}-k$, the second summand in Eqn (7.4) is

$$\sum_{k=1}^{z} \ln\frac{\frac{N}{2}+k}{\frac{N}{2}-k} = \sum_{k=1}^{z} \ln\frac{1+\frac{2k}{N}}{1-\frac{2k}{N}} = \sum_{k=1}^{z} \ln\frac{1+x}{1-x} \tag{7.5}$$

where $x = \frac{2k}{N}$.

By definition, $x \ll 1$ is always valid. We perform additional calculations to expand Eq. (7.5) and recall that

$$e^x = 1 + x + x^2 + \ldots$$

Taking into account that $x < 1$, we can limit ourselves by the two first terms, and then

$$e^x \approx 1 + x; \quad \text{i.e. } x \approx \ln(1+x)$$
$$e^{-x} \approx 1 - x; \quad \text{i.e. } -x \approx \ln(1-x)$$

This means that

$$\ln\frac{1+x}{1-x} \approx 2x; \quad \ln\frac{1+\frac{2k}{N}}{1-\frac{2k}{N}} \approx \frac{4k}{N},$$

and the relation Eq. (7.5) acquires the form

$$\sum_{k=1}^{z} \ln\frac{1+x}{1-x} = \frac{4}{N}\sum_{k=1}^{z} k = \frac{4(z+1)(z)}{2N} \approx \frac{2z^2}{N} \tag{7.6}$$

Thus, the dependence Eq. (7.3) is transformed into

$$\varphi(N;z) \approx \frac{N!}{\left(\frac{N}{2}\right)!\left(\frac{N}{2}\right)!} e^{\frac{-2z^2}{N}}$$

The obtained result can be written as follows:

$$\varphi(N;z) = \varphi(N;0)e^{\frac{-2z^2}{N}} \tag{7.7}$$

i.e. this expression means that the number of states of the system under study at any separation factor value equals the number of equilibrium states ($z = 0$) multiplied by an exponent.

The exponent coefficient can be obtained using Stirling's formula

$$n! \approx (2\pi n)^{\frac{1}{2}} n^n e^{\left(-n+\frac{1}{12n}+\ldots\right)}$$

Taking this into account,

$$\frac{N!}{\left(\frac{N}{2}\right)!\left(\frac{N}{2}\right)!} = \frac{\sqrt{2\pi N}N^N e^{-N}}{\pi N \frac{N!}{2^N} e^{-N}} = 2^N\sqrt{\frac{2}{\pi N}} \tag{7.8}$$

Hence, the relation Eqn (7.7) is

$$\varphi = 2^N\sqrt{\frac{2}{\pi N}}\, e^{\frac{-2z^2}{N}} \tag{7.9}$$

We analyse the dependence Eq. (7.9) at different ratios of the probability of the class output into fine and coarse products. We examine first the simplest case of the conditions of equilibrium distribution of a class. We analyse the obtained dependence Eq. (7.9). Its validity can be verified by summing it over all z values from $-\frac{N}{2}$ to $+\frac{N}{2}$:

$$\sum_{z=-\frac{N}{2}}^{z=+\frac{N}{2}} 2^N\sqrt{\frac{2}{\pi N}}\, e^{\frac{-2z^2}{N}},$$

Summing over all z values gives an integral

$$\int_{-\infty}^{+\infty} 2^N \sqrt{\frac{2}{\pi N}}\, e^{\frac{-2z^2}{N}} \mathrm{d}z = 2^N \sqrt{\frac{2}{\pi N}} \int_{-\infty}^{+\infty} e^{\frac{-2z^2}{N}} dz \tag{7.10}$$

We introduce a new variable

$$\frac{2z^2}{N} = y^2$$

then

$$\sqrt{\frac{2}{N}}\mathrm{d}z = \mathrm{d}y; \quad И \;\; dz = \sqrt{\frac{N}{2}}dy$$

Taking this into account, the relation under study is transformed into

$$2^N \sqrt{\frac{2}{\pi N}} \int_{-\infty}^{+\infty} e^{-y^2} \sqrt{\frac{N}{2}}\mathrm{d}y = \frac{2^N}{\sqrt{\pi}} \int_{-\infty}^{+\infty} e^{-y^2} dy$$

Because according to reference data

$$\int_{-\infty}^{+\infty} e^{-y^2} dy = \frac{\sqrt{\pi}}{2},$$

we obtain

$$\int_{-\infty}^{+\infty} 2^N \sqrt{\frac{2}{\pi N}}\, e^{\frac{-2z^2}{N}} \cdot dz = 2^N, \tag{7.11}$$

which exactly corresponds to the total number of the system states.

The distribution determined by the right-hand part of Eq. (7.9) is the Gauss distribution. It has a maximum with the center at $z = 0$.

For such curves, the mean square deviation is a measure of relative distribution width. As known, its magnitude is

$$\sigma = \sqrt{N}$$

The ratio of the mean square deviation to the maximum value is $\frac{\sqrt{N}}{N} = \frac{1}{\sqrt{N}}$. If the total number of particles constituting the system, as determined for separation conditions, equals $N \approx 10^{16}$, then the relative distribution width is on the order of 10^{-8}. It means that in this case we obtain a sharp maximum at a mean value of $z = 0$.

Its physical sense is that the separation factor attained, in fact, in a specific apparatus is not, in principle, the only possible but the most probable of all possible variants.

Here we have managed to show that the probability of this value suppresses to such extent any other conceivable distribution that it can be considered as the only possible, i.e. deterministic for the given conditions.

This explains reliably enough the constancy of the separation curve for a process with simultaneous participation of a huge number of particles. Such constancy is confirmed by all available experimental material.

We introduce another notion for the system under study. The parameter $2z$ unambiguously determines disbalance in the distribution of particles of a narrow size class between two directions.

One of principal parameters of such size class distribution is the ascending flow velocity. Obviously, the separation parameter ($2z$) value is functionally or correlationally connected with the flow velocity, but flow velocity reflects only one aspect of the process: its kinetic component. However, there is another factor in the flow, which equilibrates the kinetic component. We introduce one more parameter reflecting the potential gravitational component of this process, which should be proportional to the separation parameter, and denote it by I. We call this parameter lifting factor:

$$I = 2zc, \tag{7.12}$$

where c is the proportionality factor.

By analogy with the kinetic theory of gases, this factor should include a gravitational parameter equal, as known, to gd. Besides, to reflect the potential component, it should include the mass of a particle of a narrow size class. Finally,

$$I = -2zgdm$$

The dimension of this parameter is (kg m), i.e. the dimension of energy.

The lifting factor expresses the energy of particles disbalance. Minus appears because the gravity force is directed counter the flow. On the whole, this parameter estimates fractional extraction magnitude and direction. It has a generalising meaning for fields of various natures: centrifugal, magnetic, electric, etc. Its form for these fields differs from that of gravity field, although the method of its production is the same.

The lifting factor differential can be written as

$$dI = -2gdmdz$$

A differential of I and z or N should not be understood as an infinitesimal value, but as a number of particles several orders below these values. There is no other choice here, but instead of the differential, the sign of difference, e.g. δN can be used just as successfully.

To provide potential energy I, the measure of kinetic energy developed by the flow is determined by the minimal effort per unit area of the flow cross-section, which is denoted by f. The dimension of this parameter is (kg/m).

Potential extraction determines the deviation of particles orientation from the equilibrium one, for which $z = 0$. At the same time, one should take into account that the potential energy of a particle in a flow is gdm. In this dependence,

$$m = V(\rho - \rho_0),$$

where m is the particle mass, $V-$ its volume; ρ,ρ_0- material and medium densities. For air flows, $\frac{\rho}{\rho_0} > 1000$, and therefore, in this case one can assume that

$$m = V\rho$$

3. Generalising invariant for two-phase flows

To determine the invariant of any mass system, one needs to know how to determine the number of its admissible states. A state is considered admissible if it is compatible with characteristics of the system. An interesting situation arises at the establishment of contact between two systems. Examine the mechanism of separation factors and lifting factors exchange. Note that the main problem of statistical mechanics consists in the study of the most probable distribution between systems ensuring their mutual equilibration.

We consider a system consisting of particles of one size class and mark a part of these particles, say, by paint or by isotopes, thus obtaining two systems with different numbers of particles. We denote the numbers of particles therein by N_1 and N_2.

We determine the number of admissible states of the two systems and find the most probable configuration of a united system. We consider their behaviour in a system as follows. First, we observe particles of one class only, with N_1. Having determined the number of states of this system, we start observing the system of the second particles. After that, we examine characteristics of the system formed by particles of both classes.

Two interacting systems that exchange energies and particles eventually attain equilibrium, and their energy characteristics become the same.

We assume that the realisation of each system is determined by some fixed nonzero values of separation factors z_1 and z_2. The number of states of the first system admissible according to the separation factor is $\varphi_1(N_1; z_1)$, and each of them can be realised simultaneously with any of $\varphi_2(N_2; z_2)$ admissible self-similar states of the second system. Clearly, the total number of states in the united system is determined by the product

$$\varphi_1(N_1; z_1)\varphi_2(N_2; z_2)$$

We denote

$$z = z_1 + z_2$$

i.e.

$$z_2 = z - z_1$$

At a constant number of particles in the systems,

$$N = N_1 + N_2 = \text{const}$$

Realisation of a united system can be characterised completely enough by the product

$$\varphi_1(N_1; z_1)\varphi_2[N_2(z - z_1)],$$

i.e. through the value of one separation factor z_1.

To obtain the total number of all admissible states, it is enough to sum up the obtained expression over all z_1 values, i.e.

$$\varphi(N;z) = \sum_{z_1} \varphi_1(N_1;z_1)\cdot\varphi_2[N_2(z-z_1)]$$

As known, such a sum has a sharp maximum at a certain $z_1 \approx z_{m_1}$ value. This value of the parameter determines the most probable realisation of a united system. Then the number of states in the most probable configuration equals

$$\varphi_1(N_1;z_{m_1})\varphi_2(N_2;z-z_{m_1}) \tag{7.13}$$

Obviously, if at least in one of the two systems the number of particles is very large, this maximum is extremely sharp with respect to changes in z_1. The presence of a sharp maximum means that statistical properties of the united system are determined by a relatively small number of configurations.

Clearly, for a distribution with a sharp maximum, averaged properties of a system are precisely determined by the most probable configuration. It means that the previously found value of a physical quantity averaged over all admissible configurations can be replaced with an average over one most probable configuration only. Taking into account the previously made conclusion, we can write:

$$\varphi = \varphi_1(N_1;z_1)\varphi_2(N_2;z_2) = \varphi_1(N_1;0)\varphi_2(N_2;0)e^{\left(-\frac{2z_1^2}{N_1}-\frac{2z_2^2}{N_2}\right)} \tag{7.14}$$

We examine this relation as a function of z_1. Then Eq. (7.14) can be written as

$$\varphi = Ae^{-\left[\frac{2z_1^2}{N_1}+\frac{2(z-z_1)}{N_2}\right]} \tag{7.15}$$

Note that a certain function $\ln y(x)$ reaches its maximum at the same value of x as the function $y(x)$. We obtain from Eq. (7.15)

$$\ln\varphi = \ln A - \frac{2z_1^2}{N_1} - \frac{2z_2^2}{N_2}$$

This value has an extremum when the derivative with respect to z_1 is zero. The first derivative is

$$-\frac{4z_1}{N_1} + \frac{4(z-z_1)}{N_2} = 0$$

The second derivative

$$-4\left(\frac{1}{N_1}+\frac{1}{N_2}\right)$$

is negative, and, hence, the extremum is a maximum. Thus, the most probable configuration is that for which the following is valid:

$$\frac{z_1}{N_1} = \frac{z-z_1}{N_2} = \frac{z_2}{N_2}$$

We have obtained a very interesting relation. Two systems are in the most probable state when the relative separation factor of the first system equals the relative separation factor of the second one. This points to the fact that the most probable state of a system is established in such a way that the separation factor as if levels different amounts of particles in a flow. Thus, we have obtained a result confirming the experimentally determined invariance of fractional separation extent with respect to the initial mixture composition. This conclusion demonstrates the statistical meaning of this empirical result which was obtained long ago, but has not been explained yet.

If z_1 and z_2 are equal to z_{m_1} and z_{m_2}, respectively, for the maximum of the product under study, the obtained relation can be written as

$$\frac{z_{m_1}}{N_1} = \frac{z_{m_2}}{N_2} = \frac{z}{N} \tag{7.16}$$

Hence

$$(\varphi_1\varphi_2)_{\max} = \varphi_1(N_1; z_{m_1})\varphi_2(N_2; z - z_{m_1}) = \varphi_1(N_1; 0)\varphi_2(N_2; 0)e^{-\frac{2z^2}{N}}$$

Assume that

$$z_1 = z_{m_1} + \varepsilon; \quad z_2 = z_{m_2} - \varepsilon$$

Here ε is a measure of z_1 and z_2 deviation from their maximum values z_{m_1} and z_{m_2}. Hence, it is clear that

$$z_1^2 = z_1^2 + 2z_{m_1}\varepsilon + \varepsilon^2$$
$$z_2^2 = z_{m_2}^2 - 2z_{m_2}\varepsilon + \varepsilon^2$$

Taking this into account,

$$\varphi_1(N_1; z_1)\varphi_2(N_2; z_2) = (\varphi_1\varphi_2)_{\max} e^{\left(-\frac{4z_1\varepsilon}{N_1}\right) - \left(\frac{2\varepsilon^2}{N_1}\right) + \left(\frac{4z_2\varepsilon}{N_2}\right) - \left(\frac{2\varepsilon^2}{N_2}\right)}$$

According to the fact that

$$\frac{z_{m_1}}{N_1} = \frac{z_{m_2}}{N_2}, \tag{7.17}$$

the number of states of the configuration characterised by ε deviation from the maximum equals

$$\varphi_1(N_1; z_{m_1} + \varepsilon)\varphi_2(N_2; z_{m_2} - \varepsilon) = (\varphi_1\varphi_2)_{\max} e^{\left(-\frac{2\varepsilon^2}{N_1}\right) - \left(\frac{2\varepsilon^2}{N_2}\right)} \tag{7.18}$$

To get a deeper insight into the influence of this relation, we assume $N_1 = N_2 = 10^{10}$ and $\varepsilon = 10^6$. Hence, $\frac{\varepsilon}{N} = 10^{-4}$. For such insignificant deviation from equilibrium, we obtain $\frac{2\varepsilon^2}{N_1} = \frac{2 \cdot 10^{12}}{10^{10}} = 200$.

The product $\varphi_1\varphi_2$ constitutes a fraction equal to $e^{-400} \approx 10^{-179}$ of its maximal value. Thus, it is clear that the decrease is very strong and, hence, $\varphi_1\varphi_2$ should be a function of z_{m_1} with a very sharp peak. So, almost always, the most frequently appearing values of z_1 and z_2 are very close to z_{m_1} and z_{m_2}.

It is natural to expect significant deviations from these values in case of a small system. When considering a small system in contact with a big system, there are no theoretical difficulties. The obtained result for the number of admissible states of two systems in contact with one another can be generalised to the case of two systems taking into account the lifting factor.

Using the same reasoning as before, we find the following for the expression of self-similarity of a united system:

$$\varphi(N;I) = \sum_{I_1} \varphi_1(N_1 I_1)\varphi_2(N_2;I - I_1), \tag{7.19}$$

where the summation is performed over all the values of I_1 smaller or equal to I. Here $\varphi_1(N_1;I_1)$ is the number of admissible states of the system I at I_1. The united system configuration is determined by I_1 and I_2 values. The number of admissible states is presented by a product $\varphi_1(N_1;I_1)\varphi_2(N_2;I_2)$, and a sum over all configurations gives $\varphi(N;I)$. Define the greatest summand in this sum. It is necessary for the extremum that the respective differential was zero:

$$d\varphi = \left(\frac{\partial \varphi_1}{\partial I_1}\right)_{N_1} \varphi_2 dI_1 + \left(\frac{\partial \varphi_2}{\partial I_2}\right)_{N_2} \varphi_1 dI_2 = 0 \tag{7.20}$$

Note that $dI_1 + dI_2 = 0$. Dividing this equation by $\varphi_1\varphi_2$ and keeping in mind that $dI_1 = -dI_2$, we obtain:

$$\frac{1}{\varphi_1}\left(\frac{\partial \varphi_1}{\partial I_1}\right)_{N_1} = \frac{1}{\varphi_2}\left(\frac{\partial \varphi_2}{\partial I_2}\right)_{N_2} \tag{7.21}$$

Proceeding from the fact that

$$\frac{d}{dx}\ln y = \frac{dy}{ydx}, \tag{7.22}$$

the previous expression can be rewritten as

$$\left(\frac{\partial \ln \varphi_1}{\partial I_1}\right)_{N_1} = \left(\frac{\partial \ln \varphi_2}{\partial I_2}\right)_{N_2} \tag{7.23}$$

Here we have derived a very important relation for statistical consideration of the issue under study that we will revert to later more than once. For the time being, note the following:

First, a derivative of the logarithm of the number of self-similar states of each system with respect to lifting factor represents the most probable configuration of the system, which is the most important feature of the dependence Eq. (7.23).

Second, two systems are in equilibrium with one another when the united system is in its most probable configuration, i.e. at the maximal number of admissible states.

Third, pay attention to the quantity in the numerator of the expression Eq. (7.23):

$$H = \ln \varphi \tag{7.24}$$

The obtained expression is surprisingly simple. According to Boltzmann's classical definition, this quantity is nothing but entropy. This definition corresponds to a statement that the more admissible states a system has, the higher its entropy.

But this entropy is not derived for an ideal gas as a function of its temperature. It is derived for characterising solid phase distribution in a two-phase flow. Thus, here we introduce a notion of great physical importance into the theory of two-phase flows.

It is established that for the process under study, entropy is a function of the number of particles in a system and lifting factor, i.e.

$$H = f(N; I)$$

Entropy connection with other parameters of the process will be followed as we further expose the material.

4. Definition of chaotising factor for two-phase flows

Entropy of two-phase flows, as any other entropy, defines the uncertainty of a system state. Besides, it has some specific properties.

Let us try to define physical meaning of the expression Eq. (7.23). In essence, it is a quantity equal to entropy derivative with respect to lifting factor, which is the same for both systems, i.e.

$$\frac{\partial H}{\partial I} = \frac{1}{\chi}$$

By analogy with gas dynamics, the parameter χ plays the part of a chaotising factor of the process. Because entropy is dimensionless, the dimension of χ should be the same as that of the lifting factor (kg m), i.e. it also has the dimension of energy

$$\chi = \frac{m_0 w^2}{2} \tag{7.25}$$

where m_0 is the mass of the medium in the volume of a particle analogous to the Archimedes volume. Here the chaotising factor acquires the sense of the kinetic energy of a part of the flow equal to the volume of solid particles. When the chaotising factor of two systems that are in contact is exactly the same, the contact allows a spontaneous change in the direction of particles inside them, although the systems are in equilibrium. The number of states of the first system equals φ_1, and each of them can be realised simultaneously with any of the admissible states of the second system φ_2, i.e. the total uncertainty of two isolated systems is smaller or equal to the united system uncertainly. It means that

$$H_{\sum} \geq H_1 + H_2$$

If the entropy of the initial state H_0 is determined, the entropy of an arbitrary state is

$$H_t = H_0 + \int_0^t \frac{dI}{d\chi}$$

Usually, mass processes are basically irreversible. It is not obvious, but it is true. After all, if there exists at least one system in which entropy spontaneously decreases without applying external efforts, it can be used for reducing the entropy of some other system. It means that a spontaneous decrease in the entropy of one system leads to a spontaneous decrease in the entropy of all the systems. Hence, either all mass processes are irreversible or there are no irreversible processes at all. Usually, process irreversibility is equated to an expression similar to

$$dH \geq \frac{dI}{\chi}$$

Irrespective of positive or negative value of dI, entropy change in an irreversible process must be always positive.

Define the ratio of main parameters. In all cases, the number of the system states for a narrow size class is written as

$$\varphi = \frac{N!}{\frac{N}{2}!\frac{N}{2}!}e^{-\frac{2z^2}{N}}$$

Hence,

$$H = \ln \varphi = \ln N! - 2 \ln\frac{N}{2}! - \frac{2z^2}{N}$$

Taking into account the Stirling formula,

$$\ln n! \approx n(\ln n - 1),$$

this expression can be reduced to

$$H = N(\ln N - 1) - N\left(\ln\frac{N}{2} - 1\right) - \frac{2z^2}{N}$$

It follows that

$$H = N \ln 2 - \frac{2z^2}{N} \tag{7.26}$$

By definition of potential extraction,

$$I = -2zgdm$$

To write entropy in the form $H(N,I)$, we square both parts of this expression:

$$I^2 = 4z^2(gdm)^2$$

Hence,

$$z^2 = \frac{I^2}{4(gdm)^2}$$

Substitute this expression into Eq. (7.25)

$$H(N,I) = H(N,0) - \frac{I^2}{2(gdm)^2N}$$

From the definition of the chaotising factor, we can obtain

$$\frac{1}{\chi} = \frac{\partial H}{\partial I} = -\frac{I}{(gdm)^2 N}$$

Hence, I can be expressed through χ

$$I = -\frac{N(gdm)^2}{\chi} \tag{7.27}$$

It follows that potential extraction grows with chaotising factor.

If we examine the same system of particles in some other field (centrifugal, electric, magnetic, ultrasound, etc.), we can write in case of entropy value conservation:

$$H_0 - \frac{N(gdm)^2}{2\chi_1^2} = H_0 - \frac{N(adm)^2}{2\chi_2^2}$$

It follows that

$$\frac{g^2}{\chi_1^2} = \frac{a^2}{\chi_2^2} \quad \text{or} \quad \frac{g}{a} = \frac{\chi_1}{\chi_2}$$

where a is the acceleration of a field of other nature.

We determine the specific potential extraction value as

$$i(\chi) = \left(\frac{\partial I}{\partial \chi}\right)_{N=1} \tag{7.28}$$

This relation represents energy referred to one particle of a narrow size class. In such a form, the parameter i is constant for particles of definite size. For other particles, it is also constant but has a different value, i.e. i has a certain value independent of χ for each class.

In conclusion, we determine entropy for a two-phase flow in the regimes of pneumotransport and falling layer. In the pneumotransport regime, when all the material is lifted with the flow, $z = +\frac{N}{2}$. Taking this into account, we can write the following for a narrow size class proceeding from Eq. (7.26):

$$H = N \ln 2 - \frac{2\left(\frac{N}{2}\right)^2}{N} = N\left(\ln 2 - \frac{1}{2}\right) = 0.19315N \approx 0.2N \tag{7.29}$$

The entropy of a narrow class for a falling layer, where all solid particles precipitate against the flow, has a similar value, in this case $z = -\frac{N}{2}$.

For a polyfractional solid phase, it is insufficient to summarise Eq. (7.28) over all narrow classes to get the total entropy for its displacement in a flow. Here one more summand appears, which should not be missed: the composition entropy, which is not zero for a polyfractional mixture. Taking this into account, the total entropy can be written as follows:

$$H_0 = N_0 \ln 2 - \sum_k \frac{2z_k^2}{N_k} - \sum_k P_k \ln P_k \tag{7.30}$$

where H_0 is the total entropy of a polyfractional two-phase flow; k– the number of distinguishable narrow classes; z_k– separation factor for the k-th class; N_k– the number of particles in the k-th class; P_k– the fraction of particles of the k-th class with respect to the entire mixture of particles, or their probability.

For the pneumotransport and falling layer of a polyfractional solid phase,

$$H_0 \approx 0.2N_0 - \sum_k P_k \ln P_k \tag{7.31}$$

where $N_0 = \sum_k N_k$

CHAPTER

Substantiation of Statistical Parameters of Mass Transfer in Two-Phase Flows

8

CHAPTER OUTLINE

1. Multidimensional model of a two-phase system 121
2. Two-phase flow mobility in separation regimes 126
3. Generalising invariants for parameters characterising two-phase flows 130
4. Substantiation of entropy of two-phase flows with a polyfractional solid phase 133
5. Equilibrium state as a condition of maximal entropy 140

1. Multidimensional model of a two-phase system

Having determined a generalising parameter of a two-phase flow in the form of a potential extraction factor, we can revert to the multidimensional problem set presented at the beginning of Chapter 7.

Usually, a multidimensional system consisting of N identical parts is described by an equation with a response parameter constituting the total energy of the system. In the system under study, the potential extraction value represents the collective energy of particles.

$$I = mg(x_1 + x_2 + \ldots + x_{3N}) + \frac{1}{2m}\left(P_1^2 + P_2^2 + \ldots + P_{3N}^2\right) \tag{8.1}$$

The first summand in this expression reflects the effect of potential energy of the system on the formation of I, and the second reflects the effect of kinetic energy, because

$$\frac{1}{2}m\left(\frac{dx_i}{dt}\right)^2 = \frac{P_i^2}{2m} \tag{8.2}$$

where m is the particle mass; x_i is the particle coordinate, and P_i is the particle impulse.

In a general case, every particle is described by three coordinates and three values of impulse. Equation (8.1) is called a Hamiltonian. Recall that the main idea of statistical mechanics consists of averaging because of our unawareness of details of the behaviour of the system of particles under study. Besides, one should keep in mind that there exist some states of systems in which statistical properties stop being of interest from the standpoint of the process under study, because their probability is vanishingly low or equals unity. By way of example, refer back to Figure 4.3, where the separation curve shows particles distribution in both directions. It means that for each size within the distribution range, there are different probabilities of the output in one of directions. An averaging

Entropic Invariants of Two-Phase Flows. http://dx.doi.org/10.1016/B978-0-12-801458-5.00008-3

foundation can be formalised on the basis of a probabilistic approach. Probabilistic notions play a crucial role in statistical mechanics. It is convenient to use probabilities because they have a numerical expression between zero and unity. When the examined variables have a continuous set of values, the probability of obtaining any specified value from this continuum is zero. At the same time, a sum of probabilities must equal unity. There is nothing paradoxical about it. It is an exact analogy of the statement that a geometric point has no length, and a segment consisting of a set of points has a nonzero length. Hence, in this case, it is necessary to study the probability within an interval and not at a specified point—in other words, it is necessary to introduce probability density $P(z)$. This parameter is the probability for the z value to be confined within the interval dz.

The expression $P(z)d^nz$ is the probability for $z=(z_1, z_2, z_3 \ldots z_n)$ to be a vector comprising all continuous variables of the set under study, and d^nz is the volume of an infinitesimal cell of n-dimensional space. In this case, the condition for the sum of probabilities to equal unity is written as

$$\int_z P(z)dz = 1$$

where $P(z)$ is the probability density and z is the region of z variation in n-dimensional space. For convenience, we omit hereafter the superscript n in the differential, denoting the element of volume.

To estimate average values, the probability density is needed. It means that to estimate the average value of the $\varphi(z)$ function, it should be integrated over all z values, with the weight function equal to the probability density of z realization:

$$\langle \varphi(z) \rangle = \int_z P(z)\varphi(z)dz$$

In Chapter 7, an attempt was made to represent a solid particles continuum as a system. Although this model looks rather schematic and even, one might say, primitive, it is sufficient for obtaining rather interesting results confirmed in practice. If it is necessary to determine all possible states for a totality of particles, one can proceed differently—for example, substantiate an ensemble of systems representing copies of the system under study. Such an ensemble contains, at the same moment in time, all possible states of the system with common characteristics. We can assume that all of them have a constant number of particles N, the same energy, potential extraction, or other parameters, the values of which are specified under certain concrete conditions. To interpret an ensemble of systems, it is useful to apply n-dimensional space geometry. It is impossible to imagine such a space; it is beyond human capacity. However, an abstract analysis of such a space allows us to solve rather complicated problems with systems composed of a large number of particles. It is noteworthy that, while operating with n-dimensional space, it is useful to compare mentally its changes by analogy with the habitual three-dimensional space.

Let's look back at Figure 7.1 and examine a simplified model of the process similar to that considered in the previous section. We emphasise once more that the dynamic system characteristic of a two-phase flow consists of a large number N of particles of a narrow size class. Here we put aside the actual shape of particles and represent them as ideal balls that do not have internal degrees of freedom. This means that the state of each particle is described completely by one coordinate and one velocity component, because everything is examined as a projection onto the vertical axis.

We accept another simplification. We assume that the probability density of the system state under study, averaged over characteristic parameters of the flow, does not change with time $\left[\frac{\partial P(z)}{\partial t} = 0\right]$ and is uniform over the space $\left[\frac{\partial P(z)}{\partial x_j} = 0\right]$. These limitations should be understood not only as a steady-state process, but also as a local fulfillment of energy conservation.

After these preliminary notes, we proceed to the development of a multidimensional model of the system. In the system under study consisting of N particles, the position of each of them in space is characterised by N coordinates of their gravity centre on the vertical axis denoted by x_i, where $i = 1, 2, \ldots N$. Then, the state of such a system at a certain moment in time is determined by the values of N coordinates x_i and N corresponding velocities $\dot{x}_i$or by conjugate impulses of each particle P_i, which is practically the same.

For such a system of N particles, its general state is represented by a point in $2N$-dimensional phase space with the coordinates

$$x_1; \quad x_2, \ldots x_N; \quad P_1; \quad P_2 \cdots P_N.$$

The position of particles in the system changes with time and, respectively, the point of the phase space imaging the state of the system moves in this space, describing a line that may be called phase trajectory.

A simple global condition of a steady-state process is a certain fixed, constant value of the potential extraction. According to Eq. (8.1), we can write for the system under study

$$I = \sum_i \frac{1}{2} m v_i^2 + \sum_i m(x_i - x_0) \tag{8.3}$$

In this relation, the first term is the kinetic energy of the system and the second term is the potential energy with respect to a certain level x_0. As a result, instead of a $2N$-dimensional phase space, we obtain a $(2N + 1)$-dimensional space, where the potential extraction (I) has the meaning of a response function.

This model is much simpler than Boltzmann's model, because we fix particle distribution into two outputs only: upper and lower. Clearly, at any specified moment in time, each particle can acquire only one value of the contribution to the total separation factor. We observe a system of N particles of a certain narrow size class at subsequent moments in time t_1; t_2; $t_3 \ldots t_m$, with the number of such observations being greater than or equal to m. Naturally, at each observation, the system is in one of its states. It may happen that in n cases of observation, the system has one and the same value of separation factor equal to i. Then, the probability of such a state is

$$P(i) = \frac{n(i)}{m}$$

It follows from the definition of probability that

$$\sum_i P(i) = 1$$

The definition of the mean value of a physical quantity for a discrete system under study arises here in a natural way:

$$\langle z \rangle = \sum_i z(i)P(i) = \frac{1}{m} \sum_i z(i)n(i).$$

For a system of N particles, there are $N!$ ways of their arrangement. However, one can assume that, among these arrangements, there exist n_1 ways of obtaining z_1. From this standpoint, the systems belonging to the n_1 group become indistinguishable. It was noted that the number of states of a system or its part with the same separation factor or its values lying within a narrow interval are self-similar. Their number should be taken into account when counting the total number of states.

As a rule, the total energy of a system I is assumed to be a determining parameter for a two-phase flow model. In a stationary state of a system, where I does not depend explicitly on time, a certain point of the phase space s corresponds to each mechanical state of the system. The position of the initial point s_0 at the moment of time t_0, together with the Hamiltonian, determines the system evolution completely. The main idea of introducing the notion of an ensemble consists in considering a multitude of systems that correspond to one and the same Hamiltonian instead of one dynamic system. This is, as a rule, a very large number of systems of the same nature differing in particle velocities and in configurations of their mutual arrangement in space, so that they cover all conceivable combinations of configurations and velocities.

The choice of this multitude, or ensemble, depends on limitations imposed on the system and on initial conditions. If the initial conditions are specified unambiguously, the ensemble is concentrated in some phase space region with a clearly distinguishable boundary. Most often, initial conditions are not specified unambiguously, and therefore an ensemble is distributed, as a rule, over a broad region with fuzzy boundaries. In a general case, an ensemble is represented as a cloud of points on a response surface in the phase space. This cloud can be considered continuous, with the density

$$f(x_1; \quad x_2 \ldots x_N; \quad p_1; \quad p_2 \ldots p_N).$$

This density is normalized to unity, and therefore the distribution probability in an elementary volume in the general case is

$$dv = dx_1 \cdot dx_2 \ldots dx_N \cdot dp_1 \cdot dp_2 \ldots dp_N$$

Each specific state of the system is represented by a point in $(2N+1)$-dimensional space. All the states of the system with potential extraction I have points on a certain "surface" representing these states. States with the energy lying between I and $I+dI$ have points lying in an infinitely thin layer between I and $I+dI$ surfaces. All the states with potential extraction below I form a region limited by the surface dI. It is important to determine the size (volume) of this region. If the system is stationary, its energy I does not change with time.

For such a case, Liouville proved a theorem stating that an $(N+1)$-dimensional surface formed in this way has a limited, finite area.

For our model, Eq. (8.3) is somewhat simplified for two reasons. First, we examine a model of the process for which it is sufficient to know only one coordinate x_i and one momentum component p_i corresponding to vertical axis for each point. Second, the particle location in the volume under study is of no importance, because the contribution of particles with different potential energies into parameter

I is the same, depending on their velocity direction or on the momentum direction, which is also the same. Therefore, for each specific model under study, it is sufficient to write

$$I = \frac{c}{2m}\left(p_1^2 = p_2^2 + \ldots p_N^2\right) \tag{8.4}$$

where c is a certain proportionality factor.

Now we examine all the states of the system, the coordinates of which lie between $p_1 + dp_1$; $p_2 + dp_2$; $p_3 + dp_3 \ldots p_N + dp_N$. All coordinates in the phase space are mutually perpendicular. In this case, the volume of an elementary region is

$$dv = dp_1 \cdot dp_2 \cdot dp_3 \ldots dp_N$$

This expression represents the volume of multidimensional parallelepiped in the phase space. The entire volume of the phase space can be separated into elementary parallelepipeds. The totality of these elementary volumes and potential extraction values make it possible to determine the quantity of complexes of the system of N particles under study in a flow. An elementary layer thickness is proportional to dI. If the quantity of elementary volumes is φ, the entire volume can be represented as $\varphi \cdot dI$. We introduce the parameter φ into the entropy formula and show that the obtained expression has all the properties of entropy:

$$H = \ln \varphi$$

We obtain a relation for the dynamic entropy in the form

$$dH = \frac{dI + fdV}{\chi} \tag{8.5}$$

In addition, the relationship

$$\frac{fV}{\chi} = N$$

is substantiated. The potential extraction parameter is determined as

$$I = c_1 N \chi$$

where c_1 is the proportionality factor, taking into account the two latter relations. The dependency Eq. (8.5) can be transformed into

$$dH = \frac{dI + \frac{I}{\chi}\frac{dV}{V}}{\chi} = \frac{\frac{dI}{I} + \frac{I}{\chi}\frac{dV}{V}}{\chi}$$

The expression (8.5) represents a differential. By a proper choice of the integration constant, we can obtain

$$H = k \ln IV \tag{8.6}$$

In this expression, k is a generalised proportionality factor.

Now we check the entropy value obtained from a multidimensional model. As we know, the computation of a multidimensional space region can be simplified significantly by dividing orthogonal coordinates into several groups with mutually independent integral limits. In this case, the region size is determined as a product of a certain number of integrals. Thus, the phase space can be decomposed into any number of spaces with a lower dimension down to one particle—in other words, into N regions.

Indeed, the space region for each particle must equal V. As we know, particles occupy a negligibly small volume of this space. Therefore, each of them can occupy any position in the volume V. Thus, the phase space volume should be equal to V^N.

Now we determine the phase space volume corresponding to a thin layer between I and $I+dI$. In N-dimensional space, a hypersphere volume is proportional to r^N. Hence, we have to differentiate an expression comprising a factor $\left(\frac{2m}{c}I\right)^{\frac{N}{2}}$. Then, the desired quantity of the phase space is proportional to $\left(\frac{2m}{c}\right)^{\frac{N}{2}-1}$. Taking this into account, we can write

$$\varphi \cdot dI = \frac{2m}{c} V^N I^{\frac{N}{2}-1} dI$$

We substitute φ into the entropy equation:

$$H = \ln \varphi = \ln \frac{2m}{c} + N \ln V + N\left(\frac{1}{2} - \frac{1}{N}\right) \ln I$$

The value of N is high, and therefore the expression $\frac{1}{N}$ in brackets is negligibly small. Rejecting $\ln \frac{2m}{c}$ as something unimportant, one finally obtains

$$H = N \ln VI.$$

This expression corresponds to the dependency Eq. (8.6). Hence, entropy has been determined from the multidimensional model correctly.

2. Two-phase flow mobility in separation regimes

During the course of the analysis of main thermodynamic regularities in Chapter 1, it was concluded that the mobilities of different gases differ. The mobility value is defined as a parameter identical or proportional to heat capacity.

It is surely appropriate to bring up the question of system mobility for two-phase flows, as well. At a purely intuitive level, it is clear that this parameter is determined by the chaotizing factor value, solid-phase granulometric composition, and concentration. To determine this parameter, we examined a steady-state flow in a system shown schematically in Figure 7.1.

We assume again that the potential extraction in a system lies, as a result of fluctuations, within the range between I and $I+dI$. The point imaging this system is located in the corresponding thin layer of the phase space. We assume that in this layer, all positions of the imaging point are equiprobable, which means that layer separation into small elements of equal volume provides equal probability for a point to be located in any of these elements. Obviously, the most probable state corresponds to the largest constituent of the phase space.

Now we examine a steady-state flow in a facility shown in Figure 8.1. The two-phase flow is in the separation mode. A special feature of this facility is the presence of a partition installed along the flow.

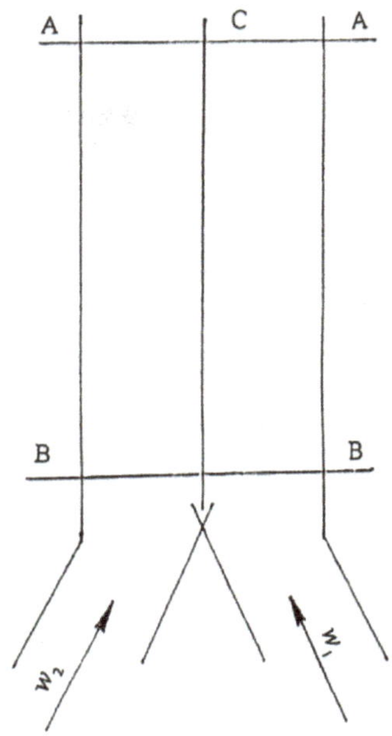

FIGURE 8.1

Scheme of a channel with a longitudinal partition.

This partition represents a grid capable of letting the carrier flow molecules pass and detain solid particles. The grid cells are so small that the particles do not stick within them. We assume that each branch of the flow contains particles of narrow size classes of different sizes x_1 and x_2, and their amounts N_1 and N_2 are also different. Besides, the velocities of the flows may differ (i.e., $w_1 \neq w_2$). At the convergence of these flows into a vertical channel, the carrier medium velocities start equalising. In one part of the channel, the velocity increases somewhat; in another, it drops in compliance with the cross-section characteristics. At that time, entropy is produced in both parts of the system. Its production stops after a steady-state flow regime is established in both parts of the channel. Assuming that the part of the channel with a stationary flow is a statistical system, the following relations are valid for it:

$$\begin{aligned} N &= N_1 + N_2 \\ I &= I_1 + I_2 \end{aligned} \tag{8.7}$$

We cannot specify exactly the total potential extraction and assume that it lies between I and $I + dI$. Indeed, if the change in I_1 for the first flow lies between I_1 and $I_1 + dI_1$, we can write the increment $I_2 + dI_2$ for the second flow. Thus, we obtain that, for system 1, the phase space region is proportional to dI_1, and for system two, to dI_2. According to the meaning of the symbol φ, the volumes of these regions can be written as

$$\varphi_1(I_1)dI_1 \quad \text{and} \quad \varphi_2(I_2)dI$$

This gives the following expression for the combined region volume:

$$\varphi_1(I_1)\phi_2(I_2)dI_1dI_2 \tag{8.8}$$

To obtain the most probable state of the combined system, it is necessary to look for the maximum of (Eq. 8.8) or its logarithm. Assuming

$$H_1 = k \log \varphi_1(I_1)$$
$$H_2 = k \log \varphi_2(I_2),$$

one must look for the maximum of the sum

$$H_1 + H_2$$

The number of admissible states of a combined system is maximal at the equality of the chaotizing factor or flow velocity in both systems.

A total change in the entropy of the combined system in a channel with a partition occurs at the expense of chaotizing factor leveling. It equals the sum of entropy changes of each part caused by equilibrating flows—in other words,

$$dH = -\frac{dI}{\chi_1} + \frac{dI}{\chi_2} = \left(\frac{1}{\chi_2} - \frac{1}{\chi_1}\right) dI$$

Hence, it is clear that when the chaotizing factor levels in both parts, entropy production becomes zero, whereas before that moment, it exceeded zero (i.e., entropy was growing). As we have shown, the most probable configuration of the combined system is the one with the maximal number of admissible states at the same chaotizing factor of both systems.

Such maximality can be determined by derivation of the product of the numbers of admissible states of separate systems with respect to independent variables characterising both systems. From the relation

$$\varphi = \varphi_1\varphi_2 = \varphi_1(N_1; I_1)\varphi_2(N - N_1; I - I_1),$$

the extremum condition is

$$d(\varphi_1\varphi_2) = \left(\frac{\partial\varphi_1}{\partial N_1} dN_1 + \frac{\partial\varphi_1}{\partial I_1} dI_1\right)\varphi_2 + \left(\frac{\partial\varphi_2}{\partial N_2} dN_2 + \frac{\partial\varphi_2}{\partial I_2} dI_2\right)\varphi_1 = 0. \tag{8.9}$$

We can write

$$dN_2 = d(N - N_1) = -dN_1$$
$$dI_2 = d(I - I_1) = -dI_1$$

Hence,

$$\frac{\partial\varphi_2}{\partial N_1} = -\frac{\partial\varphi_2}{\partial N_2}; \quad \frac{\partial\varphi_2}{\partial I_1} = -\frac{\partial\varphi_2}{\partial I_2}.$$

Dividing both parts of Eq. (8.9) by the product $\varphi_1\varphi_2$, and taking into account the resulting relations, we obtain

$$\left(\frac{\partial\varphi_1}{\varphi_1\partial N_1} - \frac{\partial\varphi_2}{\varphi_2\partial N_2}\right)dN_1 + \left(\frac{\partial\varphi_1}{\varphi_1\partial I_1} - \frac{\partial\varphi_2}{\varphi_2\partial I_2}\right)dI_1 = 0$$

This expression reflects the condition of mutual leveling or equilibrium of the systems. This dependence can be simplified somewhat:

$$\left(\frac{\partial\ln\varphi_1}{\partial N_1} - \frac{\partial\ln\varphi_2}{\partial N_2}\right)dN_1 + \left(\frac{\partial\ln\varphi_1}{\partial I_1} - \frac{\partial\ln\varphi}{\partial I_2}\right)dI_1 = 0$$

Obviously, the condition of two systems' equilibration is satisfied when the expressions in brackets acquire zero values, because the second brackets in equilibrium conditions equal zero, as established earlier. Thus, we obtain

$$\frac{\partial H_1}{\partial N_1} = \frac{\partial H_2}{\partial N_2}; \quad \frac{\partial H_1}{\partial I_1} = \frac{\partial H_2}{\partial I_2} \tag{8.10}$$

The second condition is known as $\chi_1 = \chi_2$—in other words, chaotizing factors values in both parts of the system are equalised. The first condition is new. We introduce a notation

$$\frac{\partial H}{\partial N} = \frac{\tau}{\chi} \tag{8.11}$$

where τ is a parameter representing the mobility factor, and H and N are dimensionless values; therefore, the right-hand part in Eq. (8.11) is also dimensionless.

Thus, another condition of a steady-state process is added. At the combination of two systems with the same flow velocity, the additional new condition of a stationary flow acquires the form

$$\frac{\tau_1}{\chi_1} = \frac{\tau_2}{\chi_2} \tag{8.12}$$

In other words, two systems get equilibrated when the ratios of their mobility factors to the chaotizing factor become equal. The mobility factor characterises particle behaviour under specified flow conditions. The dimension of this parameter should coincide with that of the chaotizing factor (i.e., [kgm]). This characteristic of one particle is the square of the vertical constituent of its velocity multiplied by the particle mass:

$$\tau = \frac{mv^2}{2} \tag{8.13}$$

The new parameter includes particles diffusion, as well. We now consider a nonequilibrium alternative. Let $\tau_2 > \tau_1$. At the transition of ΔN particles from system 2 into system 1, entropy change behind the partition, according to the condition Eq. (8.9), amounts to

$$dH = d(H_1 + H_2) = \left(\frac{\partial H_1}{\partial N}\right)_{I_1} \cdot dN - \left(\frac{\partial H_2}{\partial I_2}\right)_{I_2} \cdot dN = \left(-\frac{\tau_1}{\chi} + \frac{\tau_2}{\chi_2}\right) \cdot dN \tag{8.14}$$

In this expression, the statistical force of the process is $\left(-\frac{\tau_1}{\chi_1}=\frac{\tau_2}{\chi_2}\right)$, and particle redistribution represents a statistical flow. These statistical force and flow form the basis of entropy production. At the equilibrium onset—in other words, at

$$\frac{\tau_1}{\chi_1}=\frac{\tau_2}{\chi_2}$$

a directed flow of particles ceases, entropy production becomes zero, and equilibrium is established. To the first approximation, the mobility parameter determines such nonequilibrium processes as diffusion and flow velocity equalisation. Its general physical meaning is elucidated later.

3. Generalising invariants for parameters characterising two-phase flows

We can examine the interaction of solid particles in a flow in a separation regime. Gravitational fields regulate particle motion, directing them downward, whereas the chaotizing factor hinders this. We can single out a segment in the vertical channel, limiting a certain volume by two planes perpendicular to the vertical channel axis at the distance l (Figure 6.2). and can analyse the process in this segment. It does not matter how loose, solid material is fed into this segment—from below with the flow, from above counter to the flow, or through lateral holes. We assume that at a certain moment in time, there are N particles of various sizes in this segment:

$$N_n=\sum_n N_i$$

We single a specific narrow size class out of the particle flow and examine its behaviour. The flow acts on this class with a certain specific force ($f[kg/m^2]$), which allows some quantity of particles to overcome the gravity force. The medium flow rate through this volume is $V\ m^3/\text{sec}$. The chaotizing factor is χkgm. In a stationary regime, the number of particles in this volume is proportional to the volume V and specific force f, and is inversely proportional to the chaotizing factor, because the greater the latter, the faster the particles leave this volume. Hence, we can write the following for a narrow class:

$$\frac{fv}{\chi}=\sigma_i N_i \tag{8.15}$$

where σ_i is a dimensionless proportionality coefficient. This relation is derived from the process parameters. This dependence can be written for all the particles as

$$\frac{Vf}{\chi}=\sum_n \sigma_1 N_1 \tag{8.16}$$

The validity of this dependence for at least one ith size class out of the total particles distribution ensures the separation process.

The quantity f must ensure overcoming the weight of particles of a certain size within their distribution and aerodynamic resistance of the rest of the particles to the flow. This can be achieved only at the expense of flow energy equilibrating or exceeding the particles' weight. If, for all classes of particles,

$$\frac{Vf}{\chi} > \sigma_i N_i,$$

we deal with a pneumotransport regime. In the case of $\frac{Vf}{\chi} < \sigma_i N_i$ for all classes of particles, a falling or a motionless layer on a grid blown through from below is realized.

Taking all this into account, we can state that the entropy of the separation process under study is a sum of the lifting factor, volume, and particles number ($I;V;N$). It can be written as a function of many variables:

$$dH = \left(\frac{\partial H}{\partial I}\right)_{V;N} \cdot dI + \left(\frac{\partial H}{\partial V}\right)_{I;N} \cdot dV + \left(\frac{\partial H}{\partial N}\right)_{I;V} \cdot dN \tag{8.17}$$

Recall the two previously derived defining relations

$$\frac{\partial I}{\partial H} = \chi \quad \text{and} \quad \frac{\partial H}{\partial N} = \frac{\tau}{\chi}$$

Multiplying the left- and right-side parts of these expressions, we obtain

$$\frac{\partial I}{\partial N} = \tau, \quad \text{where} \quad \partial I = \tau dN$$

It follows from the first expression that $\partial I = \chi dH$.

To determine the effect of volume V and flow intensity f on the lifting factor, we examine a certain number of self-similar systems with the same entropy H, volume V, lifting factor I, and the same f.

We perform mentally a slow quasi-stationary increase in the volume of one of these systems from V to $V + \Delta V$. The change is so slow that the system remains in its initial equilibrium state (i.e., its entropy H and the number of particles N in it remain unchanged). In these conditions, such a change is reversible.

All equations of mechanics can be derived using the principle of least action (principle of the potential energy minimum). This fact was established long ago and is generally accepted. All systems in nature evolve spontaneously toward an equilibrium state, where entropy reaches its characteristic extreme values. If physical quantities acquire a minimal or maximal possible value, it means they have reached their characteristic extreme values. As we know, at constant I and V, a system evolves toward a state with maximal entropy. At the same time, by analogy with the second law of thermodynamics, a system evolves at constant H and V toward a state with minimal I. In our case, with increasing volume at a constant H, the lifting factor varies from $I(V)$ to $I(V + \Delta V)$. We expand the lifting factor into a series to the accuracy of the first-order terms with respect to ΔV:

$$I(V + \Delta V) = I(V) + \frac{\partial I(V)}{\partial V} \cdot \Delta V + \ldots$$

If, in this case, I decreases, then obviously

$$\frac{\partial I(V)}{\partial V} = -f,$$

and the general expression is written as follows:

$$I(V+\Delta V)=I(V)-\Delta Vf$$

Taking this into account, on the basis of general relation,

$$dI=\chi dH-fdV+\tau dV,$$

one can write

$$dH=\frac{dI}{\chi}+\frac{f}{\chi}dV-\frac{\tau}{\chi}dN$$

Reducing the latter expression to a common denominator, we can write

$$\chi dH=dI-\tau dN+fdV \tag{8.18}$$

Similarly, one can derive a relationship for I.

$$dI=\chi H-fdV+\tau dN$$

By analogy with thermodynamics, this dependence can be called a statistical identity for a two-phase flow. The physical meaning of this identity becomes obvious if it is written in the form

$$\chi dH+\tau dN=dI+fdV \tag{8.19}$$

The left-hand part of this identity contains the kinetic energy of the particles' flow and the right-hand side contains their potential energy. However, this identity is not a conservation law, because it does not include all flow energy and all potential energy of the solid phase. The kinetic energy refers only to a part of the flow equal to the net volume of solid particles, and not to the entire flow. Potential energy of solid particles in this expression refers only to the particles making up the unbalance. It follows from the definition of potential extraction and chaotizing factor of the flow.

In separation regimes of two-phase flows, a new, unusual property of negative production of entropy arises. When a polyfractional material enters an ascending flow, a certain ordering of particles by size occurs as a result of the stratification phenomenon. This phenomenon is opposite to mixing. If we assume that a mixture of particles can be characterised by composition entropy, the separation process leads to a decrease of this component of entropy. Concluding the chapter, we discuss one more relationship. The specific pressure of the flow can be found from the dependence

$$\frac{fV}{\chi}=\sigma N$$

Hence,

$$f=\frac{\sigma NSw^2\rho_0}{2V}$$

We denote

$$\beta=\frac{NS}{V}$$

where β is the specific volume of the solid phase in the flow and S is the volume of the isolated particle. Taking this into account,

$$f = \frac{1}{2}\sigma\beta w^2\rho_0 \tag{8.20}$$

This is analogous to the quadratic resistance law, because f is proportional to $w^2\rho_0$, which is valid from the standpoint of the physics of two-phase flows.

4. Substantiation of entropy of two-phase flows with a polyfractional solid phase

It is absolutely clear from the previous discussion that the number of particles in the system, N, is an independent statistical quantity. For a narrow class of particles, we can write the entropy of a flow using Eq. (8.19) in the form

$$dH = \frac{dI}{\chi} + \frac{f}{\chi}dV - \frac{\tau}{\chi}dN \tag{8.21}$$

For a polyfractional system consisting of k different narrow size classes containing $N_1;N_2;N_3\ldots N_k$ particles, the following expression arises for entropy, which is certainly an additive value:

$$dH = \frac{dI}{\chi} + \frac{fdV}{\chi} - \left(\frac{\tau_1}{\chi}dN_1 + \frac{\tau_2}{\chi}dN_2 + \ldots\frac{\tau_k}{\chi}dN_k\right) \tag{8.22}$$

It is unclear whether this expression is incomplete or basically wrong. The dependency Eq. (8.22) does not contain, explicitly, a very important entropy component arising at the mixing of homogeneous components—in this case, various narrow classes of particles. To clarify this issue, we represent the total entropy consisting of two parts—dynamic (H_1) and static (H_2)—as

$$dH = dH_1 + dH_2 \tag{8.23}$$

The dynamic entropy component is expressed as follows:

$$dH_1 = \frac{dI}{\chi} + \frac{fdV}{\chi} \tag{8.24}$$

The static part of this parameter, as determined for the entropy of mixing, is written as a ratio of probabilities. It is unclear which is correct:

$$dH_2 = -\sum_k P_k \ln P_k \quad \text{or} \quad H_2 = -\frac{\sum_k \tau_k N_k}{\chi} \tag{8.25}$$

By our definition, the probability of a specific size class is determined as

$$P_i = \frac{N_i}{\sum_k N_k} \tag{8.26}$$

The expression Eq. (8.26) shows a relative concentration of ith particles in a mixture—in other words,

$$P_i = \mu_i$$

Here, $\sum_k \mu_i = 1$.

To determine the connection between the relative concentration of particles and the flow parameters, we have to consider the effect of the number of particles on the character of the main regularities.

Energy, potential extraction, and entropy have the property of additivity. This property allows one to make a certain conclusion about the character of their dependence on the number of particles. Additivity of a parameter means that, at the change of the number of particles N by some factor, this parameter changes by the same factor.

We express potential extraction by the following function of entropy and volume:

$$I = Nf\left(\frac{H}{N};\frac{V}{N}\right) \tag{8.27}$$

The dependency Eq. (8.27) is the most general expression of a uniform function of the first order of $N;H;V$. Considering N as another independent variable, we can write, according to Eq. (8.21),

$$dI = \chi dH - fdV + \tau dN \tag{8.28}$$

where $\tau = \left(\frac{\partial I}{\partial N}\right)_{H;V}$ is system mobility. From this, we can show easily that

$$I = N\tau \tag{8.29}$$

Thus, the mobility of a system of solid particles in separation regimes is nothing but potential extraction per one particle. Flow mobility expressed as a function of χ and f is independent of N. We can write immediately the following expression for its differential:

$$d\tau = -hd\chi + vdf \tag{8.30}$$

where h and v are entropy and volume, respectively, per one particle.

If we divide mentally the entire volume of the system (or apparatus) into rectangular cells so that the volume of one cell admits, in a general case, at most one particle, then we can speak about a cellular model of the system under study. We assume that such division gave G cells. It was determined that solid particles occupy no more than 0.001 (or one thousandth) of the volume. Although their number is high enough, it is much less than the number of particles (i.e. $G >> N$). Placing each of N particles into one of G cells, we obtain G^N possible distributions. There are identical distributions among them, which differ only in the particles' positions in space. Recall that, according to model conditions, all the particles of a narrow class are considered identical. Therefore, the total number of permutations of N particles is $N!$ Thus, the number of complexes under such conditions is determined by the ratio

$$\varphi = \frac{G^N}{N!} \tag{8.31}$$

The entropy of such a distribution is expressed by the logarithm of Eq. (8.31):

$$H = \ln \phi = N \ln G - \ln N!$$

To expand $\ln N!$, we apply Stirling's formula in the form

$$\ln N! \approx N \ln N$$

Taking this into account,

$$H = (N \ln G - N \ln N) \tag{8.32}$$

Introducing an average value of cell filling

$$k_n = \frac{N}{G},$$

we obtain the following for the expression (8.32):

$$H = -k_n G \ln k_n \tag{8.33}$$

For the model under study, $k_n << 1$. This condition confirms that, for average cell filling values, in fact, at each moment in time there is, at most, one particle in each elementary cell. Clearly, $\sum k_n = N$. Here it is necessary to substantiate one more definition.

Conventionally, we separate a statistical system in Figure 7.1 into two parts and call the bigger one an apparatus and the smaller one a zone. By zone, we imply a part of the volume of the vertical channel having a moderate height and covering the entire cross-section of the channel. We assume the zone height is moderate, but sufficient to admit a large number of particles, and is insufficiently high for a considerable change to occur in the composition, particle concentration, and other parameters of the process. We locate the selected zone at the upper edge of the system, although it can be located in any part of the apparatus without affecting the character of the obtained results. The main limitation of zone height is determined by the fact that all particles oriented upward leave it (i.e., are extracted out of the system). The zone is in contact with the apparatus, which means they have unlimited possibilities to exchange particles, and flow velocities are equal (i.e., their chaotizing factors are the same). Besides, their mobility factors are also equal, because we imply particles of a narrow size class.

After these preliminary notes, we now study the behaviour of an individual cell. We examine the correlation between one cell and the apparatus, under the condition that particles can be both coarse and fine.

In the case of coarse particles, at most, one particle can be located in each cell. In this case, N particles must be distributed over $(G-1)$ cells. We obtain the following number of ways of distribution:

$$\varphi = \frac{(G-1)!}{N!(G-N-1)!} \tag{8.34}$$

Taking a logarithm of Eq. (8.34), neglecting unity as a result of high N and G values, and using the Stirling formula, we obtain

$$H = \{G \ln G - N \ln N - (G-N)\ln(G-N)\}$$

We introduce a parameter

$$k_n = \frac{N}{G}$$

Last, we obtain the coarse particle distribution:

$$H = G[k_n \ln k_n + (1 - k_n)\ln(1 - k_n)] \tag{8.35}$$

In the case of fine particles, any number of particles can stay in one cell. The number of ways of distributing N particles over G cells amounts to

$$\varphi = \frac{(G + N - 1)!}{(G - 1)!N!} \tag{8.36}$$

Taking a logarithm of this expression under the same conditions as that of the dependency Eq. (8.34), we finally obtain

$$H = G[(1 + k_n)\ln(1 + k_n) - k_n \ln k_n] \tag{8.37}$$

Dependencies Eqns(8.36) and (8.37) determine apparatus entropy.

Now we examine the behaviour of a separate particle connected to the apparatus. First, the cell is in the zone of the apparatus. This means that a particle located in it and oriented upward leaves the apparatus (i.e., is extracted into the fine product).

Second, we assume that all other cells belong to the apparatus. If a cell is not occupied, the probability of extraction from it is zero—in other words,

$$E = 0$$

where E is the potential extraction out of a cell. If a cell is occupied, its lifting factor has a certain value that corresponds to the probability of its orientation upward. By the definition of a large sum accepted in statistical mechanics, we can write for one cell

$$z = 1 + \lambda e^{-\frac{E}{\chi}} \tag{8.38}$$

The first term in Eq. (8.38) corresponds to the case of a nonoccupied cell and a zero lifting factor. The second term corresponds to an occupied cell with $N = 1$ and $E \neq 0$. The parameter λ is determined for the process under study as

$$\lambda = e^{\frac{\tau}{\chi}} \tag{8.39}$$

Expressions (8.38) and (8.39) are substantiated in detail in a book of the author (2010). Here, the average cell occupancy equals the ratio of the term in the large sum with $N = 1$ vs a sum of summands with $N = 0$ and $N = 1$.

$$\langle k_n \rangle = \frac{\lambda e^{-\frac{E}{\chi}}}{1 + \lambda e^{-\frac{E}{\chi}}} = \frac{1}{\lambda^{-1} e^{\frac{E}{\chi}} + 1}$$

Substituting the λ value into this dependency, we obtain

$$\langle k_n \rangle = \frac{1}{e^{\frac{E-\tau}{\chi}} + 1} \tag{8.40}$$

This dependency is characteristic of coarse particles as defined here. For fine particles, this parameter is

$$\langle k_n \rangle = \frac{1}{e^{\frac{E-\tau}{\chi}} - 1} \tag{8.41}$$

where $\langle k_n \rangle$ is a function connected with the probability of particles extraction out of a cell, with its value always varying between zero and unity.

For the flow conditions under study, the expression $\left(e^{\frac{E-\tau}{\chi}} >> 1\right)$ is always much greater than unity. Therefore, no matter whether we examine coarse or fine particles, we write

$$\langle k_n \rangle = e^{\frac{\tau-E}{\chi}} \tag{8.42}$$

The dependency Eq. (8.42) is a limiting case for the distribution of coarse and fine particles in a cellular model. It means that the total number of particles of a narrow size class equals the sum of their averaged contents in each cell. We replace the sum with an integral, keeping in mind the appearance of a proportionality factor normalizing the dimension of the obtained relation:

$$N = c\int_0^\infty e^{\frac{\tau-E}{\chi}} dE = ce^{\frac{\tau}{\chi}}\int_0^\infty dE = c\chi e^{\frac{\tau}{\chi}}$$

Hence, we obtain

$$N = c\chi e^{\frac{\tau}{\chi}} \tag{8.43}$$

The exponent in the exponential expression is dimensionless, N is also dimensionless, and thus c must be a constant with the dimension inverse to that of the chaotizing factor. Therefore, the proportionality factor can be written as

$$c = \frac{1}{gdm}$$

The parameters g, d, and m characterising a particle of a narrow size class are constant quantities.

During ordinary conditions, it is difficult to deal with the notion of the amount of particles when studying two-phase flows. In this case, the weight concentration parameter,

$$Nmg = \mu V \tag{8.44}$$

is applied, where μ is the weight concentration of particles per one cubic meter of the continuum, V is the volume of the continuum containing solid particles. From Eqns (8.43) and (8.44), we obtain

$$e^{\frac{\tau}{\chi}} = \mu\frac{Vmgd}{\chi} \tag{8.45}$$

The potential extraction parameter was found from the statistical definition of main parameters of a two-phase flow in the following form:

$$I = 2zmgd$$

Hence,

$$mgd = \frac{I}{2z}$$

Taking this into account, the relation Eq. (8.45) can be represented as

$$e^{\frac{\tau}{\chi}} = \mu \frac{VI}{2z\chi}. \tag{8.46}$$

Analysis of Eq. (8.45) from the standpoint of flow parameters leads to the relation

$$\frac{mgd}{\chi} = \frac{V_1(\rho - \rho_0)gd}{\frac{V_1 \rho_0 w^2}{2}} = 2\frac{gd}{w^2}\frac{(\rho - \rho_0)}{\rho} = 2B \tag{8.47}$$

where V_1 is the volume of a particle of a narrow size class.

Taking Eq. (8.46) into account, the dependency Eq. (8.45) is written as

$$e^{\frac{\tau}{\chi}} = \mu 2BV \tag{8.48}$$

and taking Eq. (8.45) into consideration, we obtain

$$B = \frac{I}{4z\chi} \tag{8.49}$$

Here we have established a very interesting connection between the universal invariant of two-phase flow affinization in turbulent regimes and the main statistical parameters of these flows. Note that the dependency mgd/χ contains an expression of the potential energy of a particle in the numerator and an expression of its kinetic energy in the denominator. It turns out that their ratio gives an invariant for fractional separation curves.

We return to the physical meaning of the notion of mobility. In Eq. (8.46), one parameter, flow volume, V, is not defined. For the convenience of computations, we refer all future conclusions to one cubic meter of the flow, keeping in mind that at the substitution of $V = 1$, the dimension of the unity is $[m^3]$. Taking this into account, Eq. (8.46) can be written as

$$e^{\frac{\tau}{\chi}} = \frac{\mu I}{2z\chi}$$

This expression shows the principal ratio for a narrow size class. Taking into account other classes in the flow, it is necessary to introduce indexation:

$$e^{\frac{\tau_k}{\chi}} = \frac{\mu_k I_k}{2z_k \chi}$$

Besides kth particles, the flow contains particles of other size classes, the total concentration of which is denoted by μ_0. We introduce this parameter into the numerator and denominator in the right-hand part of this expression and obtain

$$e^{\frac{\tau_k}{\chi}} = \frac{\mu_k}{\mu_0} \cdot \frac{\mu_0 I_k}{2z_k \chi}$$

For the sake of convenience, we denote the second multiplier by

$$\nu_k = \frac{\mu_0 I_k}{2z_k\chi}$$

and call this expression a discrete parameter of the kth narrow size class. Last, we can write

$$e^{\frac{\tau_k}{\chi}} = \frac{\mu_k}{\mu_0}\nu_k$$

Hence,

$$\tau_k = \chi \ln\frac{\mu_k}{\mu_0} + \chi \ln \nu_k \tag{8.50}$$

When there is only one class of particles in the flow, the ratio

$$\frac{\mu_k}{\mu_0} = 1, \quad \text{and} \quad \ln\frac{\mu_k}{\mu_0} = 0$$

In this case, the mobility parameter (τ_{k0}) acquires the form

$$\tau_{k0} = \chi \ln \nu \tag{8.51}$$

In a general case, with motion in a polyfractional mixture flow, this parameter acquires the form

$$\tau_k = \tau_{k0} + \chi \ln\frac{\mu_k}{\mu_0} \tag{8.52}$$

To clarify the statistical sense of the notion of mobility, we write

$$\tau = \frac{mv^2}{2}$$

In ideal theoretical conditions, solid-particle velocity is determined as a linear function of the mean flow velocity,

$$v = w - w_0$$

where w_0 is hovering velocity of a particle.

This dependency is derived for a separate particle only, without taking into consideration the restriction of its motion by other particles and its interaction with particles of other size classes.

Proceeding from statistical notions, the mobility parameter of particles of a narrow size class in a two-phase flow is determined by the dependency Eq. (8.52). It becomes clear that the mobility parameter of particles of a narrow size class is determined mainly by the chaotizing factor, concentration of this class in the flow and, according to Eq. (8.49), by the invariant of separation curves affinization or main statistic parameters of a two-phase flow. It has discrete values for each narrow class in the flow.

For a polyfractional mixture, the mean velocity of particles in each size class increases somewhat as a result of constrained flow conditions and the interaction of particles of various classes. However, the inequality

$$\tau < \chi$$

always holds. Now we return to dependencies Eqns (8.22) and (8.25) and examine

$$dH_2 = -\frac{\sum_k \tau_k dN_k}{\chi}$$

Taking Eq. (8.52) into account, we can write

$$H_2 = \frac{\sum_k \tau_{0k}\, dN_k}{\chi} - \sum_k dN_k \ln \mu_k \tag{8.53}$$

The first term of this expression relates to the dynamic entropy component; the second constitutes mixing entropy. In conclusion, we can write that the entropy of two-phase flows with a polyfractional solid phase is

$$dH = \frac{dI}{\chi} + \frac{fdv}{\chi} - \frac{\sum_k \tau_{0k}\, dN_k}{\chi} - \sum_k dN_k \ln P_k \tag{8.54}$$

It follows from Eq. (8.51) that the mobility parameter for a narrow size class is proportional to the chaotizing factor (χ) or velocity of the fluid flow, concentration of this class, and the discrete parameter value. The mobility of each, specific narrow size class in a polyfractional mixture is determined by the relation Eq. (8.52) , reflecting the share of this class in the total solid-phase composition. The dependency of total entropy for a polyfractional mixture can be finally written as

$$dH = \frac{dI}{\chi} + \frac{fdv}{\chi} - \frac{\sum_k \tau_{0k}\, dN_k}{\chi} - N_0 \sum_k dP_k \ln P_k \tag{8.55}$$

In this dependency, N_0 is the total number of particles in the mixture.

5. Equilibrium state as a condition of maximal entropy

The main relation connecting the distribution function ρ_n with system properties and characterising the ensemble was established by Boltzmann and was formulated in a more general form by Planck. It connects the system entropy H with the distribution function ρ_n by the equation

$$H = -k \sum_n \rho_n \ln \rho_n \tag{8.56}$$

under the condition

$$\sum_n \rho_n = 1 \tag{8.57}$$

Summation is performed over all the states constituting an ensemble (i.e., those with nonzero ρ_n). The relation Eq. (8.56) satisfies the statement that entropy H is a measure of the system disorder. In fact, if a system is in ideal order, then $\rho_i = 1$, and all other ρ_{n-1} are zero. Because $\ln 1 = 0$, the entropy H in this case is also zero. On the other hand, if the system is unordered, it can be in any of its states. The more states in has, the greater the disorder. If $\rho_n = \frac{1}{N}$ for N states and zero for all other states, then

$$H = k\sum_{n=1}^{N}\frac{1}{N}\ln\frac{1}{N} = k\ln N$$

The greater the value of N, the greater the value of H. Hence, Eq. (8.56) satisfies our ideas of entropy behaviour. One can use this dependency for finding the distribution function of various states of a two-phase system—in particular, equilibrium states. As known, entropy tends to grow until it reaches its limit value at equilibrium, obeying at the same time the limitations imposed on the system. We assume that no limitations are imposed on the distribution function ρ_n excluding the limitation Eq. (8.56), and that the number of system states is a certain finite quantity φ. Under such conditions, the maximal entropy value is determined.

The condition Eq. (8.56) means that only $(\varphi - 1)$ values of ρ_n can take on mutually independent values. One value, f_φ is connected with

$$\rho_\varphi = 1 - \sum_{n=1}\rho_n, \tag{8.58}$$

but entropy is a function of all ρ_i values and can be represented as

$$H(\rho_1;\rho_2;\rho\ldots\rho_n\ldots\rho_\varphi)$$

Here, ρ_φ can be expressed through other ρ_n, as shown by Eq. (8.58). For H to become maximal, partial derivatives of H with respect to each independent f_i should be set equal to zero. Because f_φ depends on other f_i, these equations acquire the form

$$\left.\begin{array}{c}
\dfrac{\partial H}{\partial\rho_1} + \dfrac{\partial H}{\partial\rho_\varphi}\cdot\dfrac{\partial\rho_\varphi}{\partial\rho_1} = \dfrac{\partial H}{\partial\rho_1} - \dfrac{\partial H}{\partial\rho_\varphi} = 0 \\
\dfrac{\partial H}{\partial\rho_2} - \dfrac{\partial H}{\partial\rho_\varphi} = 0 \\
\cdots\cdots\cdots\cdots\cdots\cdots\cdots\cdots\cdots\cdots \\
\dfrac{\partial H}{\partial\rho_{\varphi-1}} - \dfrac{\partial H}{\partial\rho_\varphi} = 0
\end{array}\right\} \tag{8.59}$$

The values of functions ρ satisfying these equations should satisfy the relationships Eqns (8.56) and (8.57). Obviously, the partial derivative $\frac{\partial H}{\partial\rho_\varphi}$ for these equations equals a certain value that we denote by

$$\frac{\partial H}{\partial\rho_\varphi} = -a_0$$

Then, Eq. (8.59) can be written as

$$\frac{\partial H}{\partial\rho_1} + a_0 = 0;\quad \frac{\partial H}{\partial\rho_2} + a_0 = 0;\cdots\cdots\frac{\partial H}{\partial\rho_{\varphi-1}} + a_0 = 0. \tag{8.60}$$

The quantity a_0 is determined obeying condition (8.56). This parameter (a_0) is known as the Lagrange factor. Hence, we arrive at a problem of variations calculus. We substitute Eq. (8.59) into Eq. (8.60) instead of H using Eq. (8.56), for instance, for the case of ρ_k:

$$0 = \frac{\partial}{\partial \rho_k}\left[a_0 \sum_{i=1}^{\varphi} \rho_n - k \sum_{i=1}^{\varphi} \rho_k \ln \rho_k\right] = a_0 - k \ln \rho_k - k$$

Hence, we obtain

$$\rho_k = e^{\frac{a_0}{k}-1} \tag{8.61}$$

The solution shows that, at the equilibrium state of the system, all values of ρ_k are equal, because a_0 and k are independent of the value of k. This means that for this state of the system, the values of a_0 and ρ_k follow from the requirements $\sum_{n=1}^{\varphi} \rho_k = 1$ and $\rho_k = \frac{1}{\varphi}$.
Hence,

$$H = -\sum_{k=1}^{\varphi} \frac{1}{\varphi} \ln \frac{1}{\varphi} = k \ln \varphi \tag{8.62}$$

Thus, a system limited by a finite number of states φ and having no other limitations can have maximal entropy in any of φ states with equal probability. We can see that the more states a system has, the greater its entropy. In addition, this implies two more conclusions.

First, it is proved that, in the equilibrium state, all self-similar states of a system are equiprobable. Second, we can make a seemingly senseless and paradoxical conclusion that each point of the response surface represents a maximum.

It is a very interesting result, but it calls for thorough analysis and substantiation of the notion of the equilibrium state in a two-phase flow. Namely, we need to understand what is implied by the equilibrium state for a system of particles of different sizes moving in opposite directions in a flow. We make an attempt to formulate and substantiate this complicated phenomenon for the process under study in the next chapter.

CHAPTER

Specific Properties of Entropy of Two-Phase Flows

9

CHAPTER OUTLINE

1. Composition entropy as a criterion of solid-phase separation quality 143
2. Statistical equilibrium for particle distribution in a flow 148
3. Stability of entropy and the entire mass distribution process in a two-phase flow 155

1. Composition entropy as a criterion of solid-phase separation quality

The main property of thermodynamic entropy is its growth until acquiring the maximal possible value for specified process conditions. At that point, the process evolves toward the equilibrium state.

At first glance, it is barely possible to imagine any equilibrium of a solid phase in a two-phase flow. However, the analysis of its distribution gives hope in finding nontrivial equilibrium in particle distribution connected with system entropy. The general pattern of solid-phase distribution in a flow is as follows. Light, fine particles rise together with the flow; heavy, coarse ones move counter to the flow; and intermediate classes are divided between these two directions in various ratios, with their fractional separation varying within the range of $0 \le Ff(x) \le 100\%$.

Figure 9.1 shows solid-phase composition in partial residues before they are fed into the flow in the form of the dependency $Q(x) = f(x)$. After feeding such material into the flow, and its distribution into both outputs, its composition is represented in Figure 9.1 by curves $g(x)$ and $n(x)$. Because an ideal separation process is impossible, these two curves necessarily intersect at a certain point corresponding to particles size x_0. Because of the absence of ideality, some fine (with respect to size, x_0) particles, D_c, falls out downward together with coarse particles, whereas some coarse (with respect to the same size) particles, R_f, fall out into the fine product. The content of the fine product in the upper output is D_f, whereas the content of the coarse product in the lower output is R_c. According to the graph,

$$\left.\begin{aligned} D_f + D_c &= D_s \\ R_c + R_f &= R_s \\ R_f + D_f &= \gamma_f \\ R_c + D_c &= \gamma_c \end{aligned}\right\} \tag{9.1}$$

Entropic Invariants of Two-Phase Flows. http://dx.doi.org/10.1016/B978-0-12-801458-5.00009-5

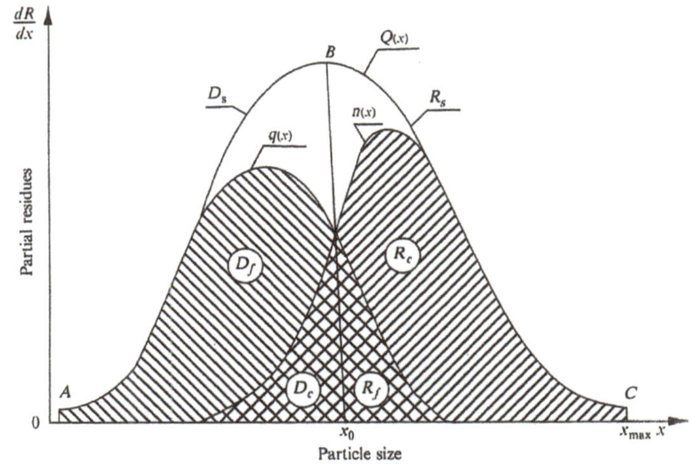

FIGURE 9.1 Bulk material distribution into two components.

where γ_f and γ_c are, respectively, the outputs of fine and coarse products. Taking the initial product composition as unity, we can write

$$R_s + D_s = 1 \tag{9.2}$$

The point x_0 is not chosen as a separation characteristic at random. It represents the size for which the separation of bulk material (Figure 9.1) is optimal, in the sense of minimal mutual contamination of both products. In this distribution, total contamination is represented by the checked area:

$$E_f = D_c + R_s \tag{9.3}$$

It is rather easy to make sure this statement is correct by moving the separation boundary to the left or to the right of x_0. In both cases, the common contamination area in the graph increases. Hence, size x_0 is optimal for this distribution from the standpoint of the separation process. To determine specific conditions for obtaining a maximal difference in the separation of products, one should minimize Eq. (9.3):

$$\frac{dE}{dx} = 0$$

We then expand this expression as

$$\frac{dE}{dx} = \frac{dD_c}{dx} + \frac{dR_f}{dx} = \frac{d\int_0^{x_0} n(x)dx}{dx} + \frac{\int_{x_0}^{x_{max}} g(x)dx}{dx}$$

As we know, a derivative of a definite integral with a variable upper limit and a constant lower limit equals the subintegral expression at the upper limit point:

$$-n(x) + g(x) = 0$$

which gives

$$n(x) = g(x) \tag{9.4}$$

Thus, we have determined the particle size value for which separation is optimal. This is a necessary, but not a sufficient, condition. If several separators achieve the same boundary x_0, they provide different effects, the values of which should be determined to make a correct choice. For this purpose, we make use of the entropy evaluation of the heterogeneity of powders composition substantiated in Chapter 5. Recall that it equals

$$H = -\sum_{i=1}^{n} P_i \ln P_i \tag{9.5}$$

As demonstrated, this function reflects impartially and unambiguously the heterogeneity of polyfractional mixture composition.

The quality criterion for the process of solid particle separation in a two-phase flow should satisfy the following boundary conditions:

1. In the case of ideal separation, this criterion must assume the maximal value.
2. In the case of separation without changes in fractional composition, this criterion must be zero.

We denote the composition entropy of the original mixture by H_s, the entropy of each component after separation by H_f and H_c, and relative output of each component by γ_f and γ_c, wherein

$$\gamma_f + \gamma_c = 1$$

As an optimization criterion, we can use a criterion that shows the extent of composition ordering completeness in separation products in comparison with the initial composition

$$E = H_s - \gamma_i\left(H_f\gamma_f + H_c\gamma_c\right)$$

According to Figure 9.1, we can write

$$H_s = -(R_s \log R_s + D_s \log D_s) \tag{9.6}$$

We determine the logarithm base in Eq. (9.6) from the condition

$$H_{s(\max)} = 1$$

Because $D_s = 1 - R_s$, we can write

$$H_s = -R_s \log R_s - (1 - R_s)\log(1 - R_s)$$

We set the derivative of the obtained expression equal to zero

$$\frac{dH_s}{dR_s} = -[1 + \log R_s - 1 - \log(1 - R_s)] = 0$$

where

$$\log R_s = \log(1 - R_s)$$

and

$$R_s = \frac{1}{2}$$

As shown, this analysis also confirms that binary mixture entropy is maximal if both components are equal.

We substitute the obtained result into Eq. (9.6):

$$H_{s(\max)} = -\frac{1}{2}\log\frac{1}{2} - \frac{1}{2}\log\frac{1}{2} = -\log\frac{1}{2} = \log 2$$

To satisfy the formulated requirement, the logarithm base should be equal to two. In this case,

$$H_{s(\max)} = 1$$

Note that at the separation of a multicomponent mixture for getting n products, the logarithm base to be chosen must be n. We check the correspondence of the expression (9.6) with the formulated boundary conditions, taking into account the notes presented earlier.

1. When obtaining ideal separation, the probabilistic characteristic of each output acquires the value equal to unity, and its entropy is $1 \cdot \log_2 1 = 0$. In this case,

$$E_{\max} = E_s$$

The idea of this relation is that the maximal possible separation efficiency is determined unambiguously by the initial mixture composition.

2. For initial mixture separation without changing the initial composition, $H_1 = H_2 = H_s$ for both separation products, and the efficiency value is

$$E = H_s - \gamma_f H_s - \gamma_c H_c = H_s - H_s\left(\gamma_f + \gamma_c\right) = 0$$

Hence, expression (9.6) corresponds to boundary conditions and determines the value of separation efficiency.

The quality criterion for estimating the perfection of a continuous separation process can be written in a general form as

$$E = -\left[(D_s \log_2 D_s + R_s \log_2 R_s) - \gamma_f\left(D_f^* \log_2 D_f^* + R_f^* \log_2 R_f^*\right) - \gamma_c\left(R_f^* \log_2 R_f^* + D_f^* \log_2 D_f^*\right)\right] \tag{9.10}$$

where D_f^*; R_f^*; D_c^*; and R_c^* are the parameters' ratios to the outputs of respective products:

$$R_f^* = \frac{R_f}{\gamma_f}; \quad D_f^* = \frac{D_f}{\gamma_f}; \quad R_c^* = \frac{R_c}{\gamma_c}; \quad D_c^* + \frac{D_c}{\gamma_c}$$

Here, $R_f^* + D_f^* = 1$ and $R_c^* + D_c^* = 1$.

In compliance with Figure 9.1, we can write several simple relations:

$$\left.\begin{aligned} \varepsilon_f &= \frac{D_f}{D_s} \\ \varepsilon_c &= \frac{R_c}{R_s} \\ k_f &= \frac{R_f}{R_s} \\ k_c &= \frac{D_c}{D_s} \end{aligned}\right\} \tag{9.11}$$

where ε_f is the fine-product extraction, ε_c is the coarse-product extraction, k_f is the fine-product contamination, and k_c is the coarse-product contamination.

In practice, it is difficult to make use of the relation (9.10) with logarithms to base 2. Therefore, we try to derive a simpler relation that establishes the optimality of the separation process from this dependency. For this purpose, we set a derivative of Eq. (9.10) equal to zero after rewriting Eq. (9.10) in the form

$$E = -\left[(R_s \log_2 R_s + D_s \log_2 D_s) - \left(R_c \log_2 \frac{R_c}{\gamma_c} + D_c \log_2 \frac{D_c}{\gamma_c} + D_f \log_2 \frac{D_f}{\gamma_f} + R_f \log_2 \frac{R_f}{\gamma_f}\right)\right] \tag{9.12}$$

After certain transformations, a derivative of Eq. (9.12) takes the form

$$\frac{dE}{dx} = Q(x)\log_2 R_s - Q(x)\log_2 D_s - n(x)\log_2 \frac{R_c}{\gamma_c} + n(x)\log_2 \frac{D_c}{\gamma_c} + q(x)\log_2 \frac{D_f}{\gamma_f} - q(x)\log_2 \frac{R_f}{\gamma_f} = 0$$

Hence,

$$Q(x)(\log_2 R_s - \log_2 D_s) = n(x)\left(\log_2 \frac{R_c}{\gamma_c} - \log_2 \frac{D_c}{\gamma_c}\right) + q(x)\left(\log_2 \frac{R_f}{\gamma_f} - \log_2 \frac{D_f}{\gamma_f}\right)$$

which leads to the expression

$$\log_2 \frac{R_s}{D_s} = \varphi_c(x)\log_2 \frac{R_c}{D_c} + \varphi_f(x)\log_2 \frac{R_f}{D_f} \tag{9.13}$$

In this expression,

$$\varphi_c(x) = \frac{n(x)}{Q(x)} \quad \text{and} \quad \varphi_f(x) = \frac{q(x)}{Q(x)}$$

It is clear that $\varphi_c(x) + \varphi_f(x) = 1$.

We transform Eq. (9.13) as follows:

$$\log_2 \frac{R_s}{D_s} - \log_2 \frac{R_c}{D_c} = \varphi_f(x)\left(\log_2 \frac{R_f}{D_f} - \log_2 \frac{R_c}{D_c}\right)$$

Hence,

$$\log_2 \frac{R_s D_c}{D_s R_c} = \varphi_f(x) \log_2 \frac{R_f D_c}{R_c D_f}$$

According to Eq. (9.12), this expression can be written as

$$\log_2 \frac{k_c}{\varepsilon_c} = \varphi_f(x) \log_2 \frac{k_f k_c}{\varepsilon_c \varepsilon_f} \tag{9.14}$$

Because, in the optimal regime,

$$\varphi_f(x) = \frac{1}{2}$$

the dependency Eq. (9.14) is transformed into

$$\frac{k_c}{\varepsilon_c} = \sqrt{\frac{k_f k_c}{\varepsilon_f \varepsilon_c}} \tag{9.15}$$

The expression (9.15) can be valid in one case only, where $\varepsilon_f = \varepsilon_c$, which automatically equalizes $k_c = k_f$, as well.

Thus, on the basis of optimality conditions, using a rather complicated entropy criterion, we have obtained a rather simple relation,

$$\varepsilon_f = \varepsilon_c \tag{9.16}$$

that reflects the optimality condition for the separation process. It is convenient to use the dependency Eq. (9.16) in industrial conditions, because from equal extractions into both outputs we can determine both the optimal separation size and the efficiency value. Surely, the greater the latter index, the greater the separation quality.

2. Statistical equilibrium for particle distribution in a flow

We recall that the entropy of a system grows until reaching the maximal possible value for specified process conditions. We have noted that the attained state of the system is an equilibrium state. Here we imply a dynamic equilibrium characterized by constancy of some parameters. In this case, the objects of the system are movable, similar to gas molecules or atoms in thermodynamics.

What is implied by equilibrium in a two-phase flow moving in the distribution regime? It turns out that for such a flow there is also a state of dynamic equilibrium leading to entropy growth up to its extreme value. A peculiar feature of a two-phase flow is that such an equilibrium is established for only one narrow size class out of the entire mixture of particles, which is divided 50/50. The rest of the particles are distributed with respect to the equilibrium class in a certain constant ratio determined by the separation curve.

As discovered, the entropy dependency for a narrow size class is

$$H(N;z) = H(N,0) - \frac{2z^2}{N}$$

As follows from this dependency, for $z = 0$, entropy acquires the maximal value. It has been noted that in a mixture of particles distributed in a two-phase flow, one can always find a certain narrow size class with $z = 0$. The maximal entropy value becomes

$$H_{\max} = H(N,0) = \frac{N!}{\frac{N}{2}!\frac{N}{2}!} \tag{9.17}$$

The condition $z = 0$ is interpreted as

$$F_f(x) = F_c(x) = 0.5$$

This is nothing but an optimality condition for the separation of a particle mixture by a fixed size, x_0, according to Figure 9.1. The remaining classes are distributed in other ratios differing from $z = 0$. Here we obtain a somewhat unusual situation. A regime of motion in which the optimal size class is divided into halves is considered as having particle flow equilibrium. Hence, the equilibrium is dynamic, and for this class the entropy is maximal. This entropy value determines the established limit it tends toward during the course of the process of solid-phase distribution between two directions.

To change the size x_0, for which it is desirable to obtain an optimal separation, one should change the velocity regime of the flow so maximal entropy corresponds to this class. For $x > x_0$, the flow velocity should be increased, whereas for $x < x_0$, it should be decreased. The formulated entropy criterion or the dependency Eq. (9.16) for a binary mixture can be used as the optimization criterion.

Here it is necessary to distinguish two aspects of the problem. It is possible to consider the entropy of a narrow-boundary size class as an optimization parameter, which allows determining the distribution regime clearly. However, it does not provide an estimation of the process completeness. To obtain a qualitative estimation, one has to act as described in the previous section in the case of initial material separation into two products. For this purpose, the entire range of particle sizes is represented as a binary mixture with respect to the boundary size. Entropic analysis of the original composition and separation products allows one to determine the distribution boundary and to evaluate unambiguously the efficiency of the performed process.

Here a new, to some extent, unexpected phenomenon is observed. At the distribution of the solid phase in the flow, the parameter characterizing mixture entropy decreases. Hence, the entropy component arising at the expense of dynamic changes inside the system can decrease despite generally accepted notions.

For the rest of known processes based on thermodynamic ideas, this parameter should grow necessarily, and this can be explained by the following reasoning. All the constituents of Eq. (8.53) reflect the process dynamics increase, except the last term, which depends on the mixture composition. The process parameters do not appear in the latter; it just reflects the granulometric composition of the solid phase, and this part of entropy can decrease at its ordering. This does not mean, however, that the total entropic process decreases in this case.

Another interesting aspect of this parameter is that it has allowed, for the first time in engineering practice, an unambiguous estimation of the quality of original material separation into m products for $m > 2$.

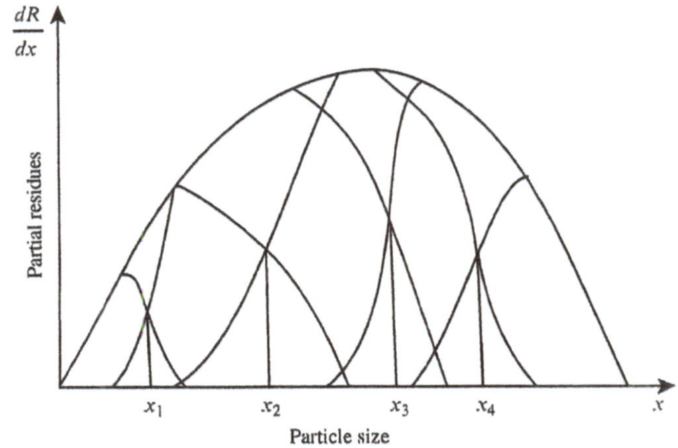

FIGURE 9.2 Schematic diagram of multiproduct separation.

Figure 9.2 shows a schematic diagram of multiproduct separation. In the case of m boundaries, $m+1$ products are obtained. Total contamination at the expense of adjacent products only in the case of multiproduct separation according to Figure 9.2 can be defined as

$$A = (D_1 + R_1) + (D_2 + R_2) + (D_3 + R_3) + \ldots + (D_m + R_m)$$

The optimization condition can be obtained here, because in the case of binary separation, by minimizing the absolute value of the contamination of all the products,

$$\frac{dA}{dx} = \frac{d(D_1 + R_1)}{dx} + \frac{d(D_2 + R_2)}{dx} + \ldots + \frac{d(D_m + R_m)}{dx} = 0 \tag{9.18}$$

This equality can be expanded correctly only by setting the contents of each set of brackets equal to zero—in other words, in each separation boundary, the following should hold:

$$\begin{aligned} g(x_1) &= n(x_1) \\ g(x_2) &= n(x_2) \\ g(x_3) &= n(x_3) \\ &\cdots\cdots\cdots \\ &\cdots\cdots\cdots \\ g(x_m) &= n(x_m) \end{aligned} \tag{9.19}$$

The same result can be obtained in the case of each product contamination, not only with the material of adjacent fractions. In this case, R_1, R_2, ...R_n denote total contamination of the corresponding material, which means that at each separation boundary, a narrow class should be divided in half.

The zero in Eq. (9.18) can also be obtained by shifting the boundary to the left in one case, and to the right of the conditions (Eq. (9.19)) in another. However, as demonstrated for binary mixtures, this leads to an increase in the total contamination only. Therefore, the conditions in Eq. (9.19) for a multiproduct process are the only correct ones.

For our research, a shelf classifier was chosen. It consists of seven stages, with initial feeding to the second stage counting from the top down. The experiments were performed on ground quartzite with particles within a broad size range. The main characteristics of this material are shown in Table 9.1.

Table 9.1 Original Material Characteristics

Mesh Size, x, μm	Average Narrow Class Size, $\bar{x}$, μm	Partial Residues on Sieves, r, %	Product Boundaries, μm	Product Notation	Each Product Content in the Initial One, %
1	2	3	4	5	6
2800	3000	0.8		G	3.02
2500	2650	2.24	+2150		
1800	2150	6.73	−2150	F	6.73
			+1400		
1000	1400	12.07	−1400	E	23.28
750	825	11.21	+690		
630	690	10.7	−690	D	26.9
400	515	16.2	+357.5		
315	357.5	13.73	−357.5	C	23.93
200	257.5	10.2	+180		
160	180	6.1	−180		
100	130	3.31	+71.5	B	11.84
80	90	2.43			
63	71.5	2.03			
50	56.5	1.28	−71.5	A	4.3
40	45	0.7			
0	20	0.29			

The value of the narrow size class was determined as a simple average of mesh dimensions of two adjacent sieves. The entire size range was divided by six boundaries into seven products. Table 9.1 shows the quantity of each product in the original material and the narrow class to be separated in an optimal way to obtain minimal mutual contamination of products during each separation act.

On the basis of Table 9.1, the entropy value for the original composition was obtained using Eq. (9.5):

$$H_s = +1.71$$

A preliminary set of experiments allowed us to determine optimal air flow velocities per total cross-section of the classifier, which ensures optimal separation of narrow classes given in the third column of Table 9.1. It was established that, in this apparatus, velocities were 0.95 m/s for 71.5-μm class, 1.34 m/s for 180 μm, 1.89 m/s for 357.5 μm, 2.62 m/s for 690 μm, 3.74 m/s for 1400 μm, and 4.63 m/s for 2150 μm.

Three groups of experiments with successive separation of the original material into the seven mentioned products were carried out. In the first group, the finest product, A, was separated the first in an optimal way, then product B was separated out of the residue in an optimal way, too, then product C, and so forth, up to product G.

The results of this research are summarized in Table 9.2, which shows the output of each product. Consider product D, for example. Its output with respect to initial composition is $\gamma_D = 26.705\%$. In

Table 9.2 Results of Multiproduct Separation of Original Material on a Fine-to-Coarse Basis

Separation Stage and Flow Velocity	Original Material, 100%	
	Fine Product	Coarse Product
1 $w = 0.95$ m/s	Product A $\gamma_A = 3.545\%$ $\gamma_z = 2.065\%$ $\gamma_p = 1.48\%$	Residue $\gamma = 96.455\%$
2 $w = 1.34$ m/s	Product B $\gamma_B = 12.285\%$ $\gamma_z = 6.115\%$ $\gamma_p = 6.17\%$	Residue $\gamma = 84.17\%$
3 $w = 1.89$ m/s	Product C $\gamma_C = 20.96\%$ $\gamma_Z = 15.05\%$ $\gamma_P = 5.945\%$	Residue $\gamma = 63.21\%$
4 $w = 2.62$ m/s	Product D $\gamma_D = 26.705\%$ $\gamma_z = 17.81\%$ $\gamma_p = 8.895\%$	Residue $\gamma = 36.5\%$
5 $w = 3.74$ m/s	Product E $\gamma_E = 21.14\%$ $\gamma_z = 14.903\%$ $\gamma_p = 6.237\%$	Residue $\gamma = 15.365\%$
6 $w = 4.63$ m/s	Product F $\gamma_F = 10.01\%$ $\gamma_z = 3.0\%$ $\gamma_p = 7.01\%$	Product G $\gamma_F = 5.362\%$ $\gamma_z = 2.242\%$ $\gamma_p = 3.12\%$

this product, target fractions (i.e., particles within the size range from 357.5–690 μm [$\gamma_z = 17.81\%$]), constitute 17.81% of the indicated amount. In addition, in this product there are particles coarser than 690 μm and finer than 357.5 μm. Their total amount represents the contamination of this product. For product D, their amount is 8.895% and is denoted by $\gamma_p = 8.895\%$.

The entropy of each product is determined on the basis of the contents of the target fractions (γ_z) and the contamination (γ_p) in them. The product is considered independently (i.e., the sum of these two constituents is taken as unity or 100%). In this case,

$$z = \frac{17.81}{26.705} = 66.69\%$$

and

$$p = \frac{8.895}{26.705} = 33.31\%$$

Table 9.3 Results of Multiproduct Separation of Original Material on a Coarse-to-Fine Basis

Separation Stage, Flow Velocity	Original Material, 100%	
	Fine Product	Coarse Product
1 $w = 4.63$ m/s	Residue $\gamma = 94.02\%$	Product G $\gamma_G = 5.98\%$ $\gamma_z = 2.375\%$ $\gamma_p = 3.605\%$
2 $w = 3.74$ m/s	Residue $\gamma = 83.795\%$	Product F $\gamma_F = 10.225\%$ $\gamma_z = 3.0\%$ $\gamma_p = 7.225\%$
3 $w = 2.62$ m/s	Residue $\gamma = 61.74\%$	Product E $\gamma_E = 22.056\%$ $\gamma_z = 14.868\%$ $\gamma_p = 7.188\%$
4 $w = 1.89$ m/s	Residue $\gamma = 35.04\%$	Product D $\gamma_D = 26.7\%$ $\gamma_z = 17.826\%$ $\gamma_p = 8.874\%$
5 $w = 1.34$ m/s	Residue $\gamma = 14.495\%$	Product C $\gamma_C = 20.545\%$ $\gamma_z = 15.015\%$ $\gamma_p = 5.53\%$
6 $w = 0.95$ m/s	Residue A $\gamma_A = 3.453\%$ $\gamma_z = 3.0\%$ $\gamma_p = 0.453\%$	Product B $\gamma_B = 11.049\%$ $\gamma_z = 8.42\%$ $\gamma_p = 2.629\%$

In this case, composition entropy for each specific product D is written as

$$H_D = -(0.6669 \ln 0.6669 + 0.3331 \ln 0.3331) = +0.6362$$

Composition entropies for all other products were calculated in a similar way:

$$H_A = +0.6795; \quad H_B = +0.696; \quad H_C = 0.5958; \quad H_E = +0.606;$$
$$H_F = +0.6096; \quad H_G = +0.6796$$

Then, the total efficiency of multiproduct separation was determined using Eq.n (9.5):

$$E_1 = 0.632$$

In the second group of experiments, each product was separated from the original material vice versa—on a coarse-to-fine basis—with product G first, then F, and so on, to product A. The results of this group of experiments are presented in Table 9.3. On the basis of Table 9.3 data, we was established for this case that $H_G = +0.6718$:

$$H_F = -0.6052; \quad H_E = -0.6313; \quad H_D = -0.6361; \quad H_C = -0.5756;$$
$$H_B = -0.549; \quad H_A = +0.388.$$

Using Eq. (9.5), the total efficiency of such multiproduct separation is

$$E_{II} = 0.6466$$

In the third group of experiments, the principle of efficiency maximization at each separation step was applied. First, a narrow size class was determined in the original composition, for which the contents of fine and coarse products are the closest to the same ratio. This is a narrow class with an average particle size of $\bar{x} = 690$ μm. With respect to this class, total residues $R_s = 43.73\%$, and total passes $D_s = 56.27\%$. Therefore, the first separation stage was carried out at the airflow velocity of $w = 2.62$ m/s. Here, the output into the fine product was 62.04% of all original material, and the output into the coarse product was 37.96%. In the fine-product output, the class with the average particle size

Table 9.4 Results of Multiproduct Separation of the Original Material on the Basis of Maximal Efficiency Principle during Each Separation Step

Separation Stage Airflow Velocity	Original Material, 100%	
	Fine Product	**Coarse Product**
1 $w = 2.62$ m/s	Fine product I $\gamma = 62.04\%$	Coarse product I $\gamma = 37.96\%$
2 $w = 1.89$ m/s from fine product I	Fine product II $\gamma = 35.05\%$	Product D $\gamma_D = 27\%$ $\gamma_z = 17.83\%$ $\gamma_p = 9.13\%$
3 $w = 3.74$ m/s from coarse product I	Product E $\gamma_E = 22.6\%$ $\gamma_z = 14.97\%$ $\gamma_p = 7.63\%$	Coarse product III $\gamma = 15.34\%$
4 $w = 1.34$ m/s from fine product II	Fine product IV $\gamma = 14.97\%$	Product C $\gamma_C = 20.079\%$ $\gamma_z = 15.025\%$ $\gamma_p = 5.054\%$
5 $w = 4.63$ m/s from coarse product III	Product F $\gamma_F = 10.13\%$ $\gamma_z = 3\%$ $\gamma_p = 7.13\%$	Product G $\gamma_G = 5.8\%$ $\gamma_z = 2.11\%$ $\gamma_p = 3.69\%$
6 $w = 0.95$ m/s from fine product IV	Product A $\gamma_A = 3.6\%$ $\gamma_z = 3.08\%$ $\gamma_p = 0.52\%$	Product B $\gamma_B = 11.37\%$ $\gamma_z = 10.153\%$ $\gamma_p = 1.217\%$

$\bar{x} = 357.5$ μm was the closest to 50% composition. With respect to this class, $R_s = 57.5\%$ and $D_s = 42.5\%$. In the coarse-product output, an analogous size class was that of $\bar{x} = 1400$ μm. With respect to this class, $R_c = 31\%$ and $D_c = 69\%$.

The second separation was performed with fine-product output at a velocity of $w = 1.89$ m/s, which corresponds to the optimal separation for class 357.5 μm. The coarse material in this separation was product D, which contained 27% of the original material.

The third separation in this group of experiments was performed with the coarse-product output remaining after the first experiment at an airflow velocity of $w = 3.74$ m/s, which is optimal for the narrow class with an average particle size $\bar{x} = 1400$ μm. The fine material from this separation was product E, which contained 27% of the original material. The separation was completed in a similar way.

A representation of this separation is shown in Table 9.4. On the basis of the data in Table 9.4, composition entropies of each product are estimated as follows:

$$H_D = +0.641; \quad H_E = +0.6396; \quad H_C = +0.6183; \quad H_F = +0.607;$$

$$H_G = +0.6555; \quad H_A = +0.412; \quad H_B = +0.3368$$

Using Eq. (9.5), we see that, in this case, the total efficiency of separation into seven products is

$$E_{III} = 0.6526,$$

which exceeds the total separation efficiency value in the two previous cases.

3. Stability of entropy and the entire mass distribution process in a two-phase flow

Filming of a two-phase flow in the conditions of solid-phase distribution shows a developed chaos in particle motion. This chaos is accompanied by the displacement of particles of various classes upward and downward, and to the left and to the right. They take part in vortical motion and create permanently forming and decaying aggregates. Particles of various sizes collide with each other, channel walls, and internal elements. Nevertheless, separation curves resulting from such a process have a markedly deterministic stable character. This fact is elucidated in detail in Chapter 6. It seems that each size class of particles behaves independently in this chaotic motion, regardless of the behaviour of all other size classes, although the particles undoubtedly interact.

This phenomenon can be explained only by analogy with Dalton's law for a mixture of gases. According to this law, molecules of all components interact, and the pressure on the chamber walls is formed by independent pressures of each component, which it could exert if it occupied the entire volume. It has been proved experimentally that a universal separation curve obtained on a specific apparatus is practically absolutely stable. It remains constant with changes in granulometric composition of the solid phase, airflow velocity within the limits of turbulent mode, boundary separation size, and even material concentration in a flow within certain limits. Naturally, a question arises: What ensures that such a deterministic stability of separation results in a practically completely chaotic process with the participation of an immense number ($10^{10} - 10^{11}$) of particles of various sizes, densities, and irregular shapes? Clearly, the main parameters of the process are subject to permanent fluctuations. Nonuniformity of local concentration and flow velocity, compositions of particles of various classes, flow pressure, and so on, take place.

To look into all this, we examine a defining relation for a critical flow system with a polyfractional solid phase:

$$dI = \chi dH - fdV + \sum_k \tau_k dN_k \tag{9.20}$$

In this expression, entropy H and potential extraction I are functions of state, whereas χ, τ, f, and N are active parameters of the system.

In a classic case, entropy of a system is formed under two different impacts on the system:

$$dH = d_eH + d_iH$$

where d_eH is a part of entropy resulting from the impact of the external (with respect to the system) medium, d_iH is a part of entropy resulting from irreversible internal changes in the system during the course of the process.

Here, a part of entropy d_eH can grow or be zero or, in certain conditions, can even take up a negative value. As for d_iH, it is assumed that it always grows (i.e., $d_iH > 0$). Indeed, the change in both constituents in all cases is such that $d_eH + d_iH > 0$.

Irreversible processes can be described in terms of mass forces and mass flows, with the flows arising as a result of forces. For example, concentration or flow velocity gradients in various parts of a system are mass forces. In this case, a flow can be represented as a number of particles dN displaced (transformed) between these parts during time dt. In this case, the entropy change is represented as

$$d_iH = FdN$$

where F is the mass force. For open systems,

$$d_eH = \frac{dI + fdV}{\chi}$$

The quantity d_eH implies the change in entropy caused by a flow of particles entering and leaving the system in opposite directions. Here, the mobility factor plays a defining role.

The total entropy for a narrow size class is written as

$$dH = \frac{dI + fdV}{\chi} - \frac{\tau dN}{\chi}$$

Obviously, in this dependency

$$d_{i1}H = -\frac{\tau dN}{\chi}$$

The total number of particles of a narrow class in the process is determined by their inlet into the system, outlet out of the system, and internal displacements, which means we can write

$$dN = d_eN + d_iN$$

As shown, the potential extraction for a narrow size class is determined by the number of particles and the chaotizing factor value:

$$I(\chi;N) = NI(\chi)$$

where $I(\chi)$ is potential extraction probability for one particle.

By analogy, we can write for a polyfractional mixture

$$I(\chi;N) = \sum_i I_i(\chi;N) = \sum_i N_i I_i(\chi)$$

where $i = 1, 2, 3\ldots$, i is the number of a component of a polyfractional mixture of particles. In this case, entropy is written as

$$d_e H = \frac{dI}{\chi} + \frac{fdV}{\chi}$$

and internal entropy is written as

$$d_{i1} H = -\sum_i \frac{\tau_i d_i N}{\chi}$$

It is accepted that its change is always positive. If we add these two expressions, the total entropy is positive:

$$dH = d_e H + d_i H = \frac{dI}{\chi} + \frac{fdV}{\chi} - \sum_i \frac{\tau_i dN}{\chi} > 0$$

In some processes, such as separation, certain phenomena arise that make an additional contribution to entropy, which leads to a decrease in the composition uncertainty in the separation process and in the products in comparison with the original composition.

A change in granulometric composition in the distribution process leads to a decrease in composition uncertainty in the separation process and in the products in comparison with the original composition, because a certain composition ordering occurs in each direction of the solid-phase motion. Namely, light, fine particles rise primarily, whereas coarse, heavy ones fall. The following always holds:

$$d_{i2} H < 0$$

This decrease takes place within the system. Indeed, a specific form of equilibrium sets in, in which the boundary class is extracted equally into both products (i.e., $F_f(x) = F_c(x) = 50\%$).

Hence, the boundary class entropy must be taken as the basis for process stability analysis. Clearly, the equilibrium state in a real process is violated permanently by fluctuations of various separation parameters such as χ, τ, f, and N. They fluctuate around their average values. Fluctuations of each parameter cause changes in equilibrium entropy. Surely, these changes can decrease the latter only:

$$\Delta H_0 < 0$$

Nonequilibrium processes occurring in a system level such fluctuations and return entropy to its original value, which is extreme for these conditions. Otherwise, the system loses its stability, which does not occur in real practice. Thus, a system is stable with respect to fluctuations when entropy changes decrease the latter. The problem consists of determining the probability of fluctuations of a concrete parameter and the conditions in which they become essential.

To obtain the probability of fluctuations, one needs to obtain a related entropy change. Thus, the main problem is reduced to the derivation of the ΔH connection with fluctuations in process parameters such as $\delta\chi$, $\delta\tau$, δf, and δN. In a general case, entropy can be expanded as

$$H = H_0 + \sigma H + \frac{1}{2}\sigma^2 H + \cdots\cdots, \tag{9.21}$$

where σH is a term of the first infinitesimal order comprising σI, $\sigma\chi$, σN, and so on, of the first infinitesimal order; $\sigma^2 H$ is a term of the second infinitesimal order comprising $\sigma^2 I$,$\sigma^2\chi$, $\sigma^2 N$, and so on, of the second infinitesimal order; and H_0 is steady-state entropy.

First, we examine the simplest situation. Assume that the fluctuation has occurred in a small part of the limited system or in the zone. Its consequence is N, χ, f flow from one part of the system into another. Entropy of the system under study can be expressed by a sum

$$H = H_1 + H_2, \tag{9.22}$$

where 1 is the part of the system where the fluctuation occurred and 2 is the rest of the system:

$$H_1 = f(N_1; \chi_1; \tau_1; f_1 \text{ etc.})$$

and

$$H_2 = \varphi(N_2; \chi_2; \tau_2; f_2 \text{ etc.})$$

We then examine process stability vs flow velocity fluctuations (w, χ). We apply Eq. (9.21) and express entropy deviation from the equilibrium state at the expansion into the Taylor series as follows:

$$H - H_0 = \Delta H = \frac{\partial H_1}{\partial I_1}\sigma I_1 + \frac{\partial H_2}{\partial I_2}\sigma I_2 + \frac{\partial^2 H_1}{\partial I_1^2}\frac{\sigma I_1^2}{2} + \frac{\partial^2 H_2}{\partial I_2^2}\frac{\sigma I_2^2}{2} + \cdots\cdots \tag{9.23}$$

Terms of higher infinitesimal order in this expansion can be neglected. Note that all derivatives in Eq. (9.23) refer to the equilibrium state. Because the potential extraction (or extraction from the zone) in a steady state remains constant,

$$\sigma I_1 = -\sigma I_2 = \sigma I$$

On the other hand, it was established that

$$\left(\frac{\partial H}{\partial I}\right)_{V,N} = \frac{1}{\chi}$$

Taking this into account, dependency Eq. (9.23) can be rewritten as

$$\Delta H = \left(\frac{1}{\chi_1} - \frac{1}{\chi_2}\right)\sigma I + \left(\frac{\partial}{\partial I_1}\frac{1}{\chi_1} + \frac{\partial}{\partial I_2}\frac{1}{\chi_2}\right)\frac{\sigma^2 I}{2} \tag{9.24}$$

Now we can determine the entropy deviation of the first and second infinitesimal order σH and $\frac{\sigma^2 H}{2}$:

$$\sigma H = \left(\frac{1}{\chi_1} - \frac{1}{\chi_2}\right)\sigma I$$

and

$$\sigma^2 H = \left(\frac{\partial}{\partial I_1}\frac{1}{\chi_1} + \frac{\partial}{\partial I_2}\frac{1}{\chi_2}\right)\frac{\sigma^2 I}{2}$$

In the equilibrium state, $\chi_1 = \chi_2$, and hence $\sigma H = 0$.

It follows that only fluctuations of the second infinitesimal order introduce changes into the entropy. It has been shown that

$$\frac{\partial}{\partial I}\frac{1}{\chi} = \frac{1}{\chi^2}\frac{\partial \chi}{\partial I} = \frac{1}{\chi^2}\frac{1}{i}$$

where $i = \frac{\partial I}{\partial \chi}$ is a specific potential extraction. Hence, we can write

$$\sigma I = i\sigma\chi$$

Taking this into account,

$$\sigma^2 H = -\frac{i(\sigma\chi)^2}{\chi^2} < 0$$

To satisfy this condition, the specific extraction should be positive (i.e., $i > 0$). Otherwise, the system loses stability, and hence, in separation processes, $i > 0$ always holds.

Consider the case of fluctuations in the number of particles in a certain part of the system's cross-section. As in the first case, for such fluctuations we can write

$$H - H_0 = \Delta H = \frac{\partial H_1}{\partial N_1}\sigma N_1 + \frac{\partial H_2}{\partial N_2}\sigma N_2 + \frac{\partial^2 H_1}{\partial N_1^2}\frac{\sigma^2 N_1}{2} + \frac{\partial^2 H_2}{\partial N_2^2}\frac{\sigma^2 N_2}{2} \tag{9.25}$$

Note that the diffusion of particles from one part of the system to another is $\sigma N_1 = -\sigma N_2 = \sigma N$, and

$$\frac{\partial H}{\partial N} = -\frac{\tau}{\chi}$$

Then, taking these observations into account, Eq. (9.25) can be rewritten as

$$\Delta H = \sigma H + \frac{\sigma^2 H}{2} = \left(\frac{\tau_2}{\chi} - \frac{\tau_1}{\chi}\right)\sigma N - \left(\frac{\partial}{\partial N_1}\frac{\tau_1}{\chi} + \frac{\partial}{\partial N_2}\frac{\tau_2}{\chi}\right)\frac{\sigma^2 N}{2}$$

Because the derivatives are taken for the equilibrium state, $\tau_1 = \tau_2$, and thus the first summand vanishes and the entropy change dependence on the whole becomes

$$\frac{\sigma^2 H}{2} = -\left(\frac{\partial}{\partial N}\frac{\tau}{\chi}\right)\frac{\sigma^2 N}{2}$$

If this condition is satisfied, the system is stable against diffusive particles exchange. The influence of other parameters on entropy stability can be also determined in this way.

In fact, the character of fluctuation effects in real processes is much more complicated. Thus, changes in the chaotizing factor χ change not only the carrier medium flow, but also other flows, such as that of solid particles N, as well as concentration μ and pressure f in the flow. In turn, a change in any of these parameters, such as concentration, may affect χ, I, N, and f. Thus, a cross-effect occurs. The most obvious example is cross-diffusion, when the concentration gradient of particles of one size causes a diffusive flow of particles of another size.

One can easily conceive that diffusion is a permanent process during the course of separation, because particles of different classes, surely, are not distributed uniformly within the apparatus volume. Meanwhile, permanent disturbances depend on other parameters, too, and it seems extremely complicated to determine their mutual effect on the entropy value. Therefore, for a pure evaluation of all combined disturbances, a relation for thermodynamic entropy fluctuation suggested by Prigogine can be used, which can be written, by analogy, for a critical flow in the form

$$\Delta H_i = -\frac{mc}{2} \tag{9.26}$$

where ΔH_i is average entropy deviation, m is the number of independent variables, and c is a certain constant.

The simplicity of this relation is unique and explicable. The point is that entropy deviation from its extreme value is really affected by fluctuations of the second infinitesimal order. Evidently, it does not matter which parameter fluctuates; their impact is small and can be summed up in a simple way. Each independent parameter introduces a quantity equal to $-\frac{c}{2}$ into ΔH_i.

It should be reemphasized that all derivations performed in this chapter refer to N particles of the boundary separation size. Entropic stability refers to that class of particles only. The most interesting point is that, with respect to this class, all other particles classes also behave steadily, because the separation curve remains unchanged. We could confine ourselves to this study of stability in a critical flow. However, the question of how other classes of particles, the entropy of which does not reach extreme values, acquire stability remains unsolved. Probably, some other parameters ensure their stability, and this calls for further investigation.

CHAPTER

Invariants for Separation Curves 10

CHAPTER OUTLINE

1. Regimes of solid-phase particle motion in flows 161
2. Particle extraction in two-phase flow separation regimes 167

1. Regimes of solid-phase particle motion in flows

Three flow regimes can be distinguished: turbulent, laminar, and transient. Each regime affects the solid phase in a special way. However, solid particles are characterised by their own flow regimes around them that have similar characteristics: turbulent, laminar, and transient. The relations between the flow regimes in general and flow around particles predetermines the regularity of solid-particle interaction with the flow. The point is that, for instance, a laminar flow regime around the particles can take place in turbulent and transient flow regimes. As we know, all three cases of flow around the particles are possible in the transient regime. This complicated knotty problem should be clarified; otherwise, it is impossible to explain the main regularities of mass transfer in two-phase flows.

As we know, the continuum regime motion is determined by the Reynolds number (Re) calculated for a channel. Regimes have been studied and described in detail. In addition, flow regimes around solid particles placed in the flow (Re_0) have also been distinguished. There is an immense difference between threshold values of Re and Re_0. Clearly, if the flow regime around a particle is turbulent, its behaviour in a turbulent flow is unambiguous. The question is how such a particle, with turbulent flow around it, behaves in a laminar flow and, to the contrary, how a particle, with laminar flow around it, behaves in a turbulent flow.

We analyse all possible cases and attempt to develop methods of ascertaining drag coefficients and other determining parameters of mass transfer in separation regimes of two-phase flows.

Flow velocities in which the entire solid phase ascends ensure a so-called transport regime. Its velocity should be such that even the coarsest particles cannot precipitate against the current. The minimal value of this velocity is called critical velocity of pneumo- or hydrotransport, depending on the continuous medium: air or water.

The opposite flow regime, with the formation of a so-called falling layer, is also applied in technology. In this regime, all the particles precipitate against the fluid flow. In this case, the maximal ascending flow velocity, at which even the finest solid-phase particles are not transported by the flow, is called the critical velocity of a falling layer.

Entropic Invariants of Two-Phase Flows. http://dx.doi.org/10.1016/B978-0-12-801458-5.00010-1

The relation between these critical velocities is of certain theoretical and practical interest from the standpoint of the character of mass exchange processes within their range, in which processes used widely in today's industrial practises such as boiling bed and powder separation by particle size or density are realised.

To analyse the behaviour of a two-phase flow, we examine a simplified model with a vertical pipe and a circular cross-section with the radius R, assuming that particles of each narrow size class are spherical.

According to the Newton–Rittinger law, the dynamic effect of flow on an isolated particle in a turbulent regime is

$$F_r = \lambda \frac{\pi d^2}{4} \rho_0 \frac{(u_r - v_r)^2}{2}, \tag{10.1}$$

where λ is the particle resistance coefficient, d is the particle diameter, ρ_0 is the fluid density, r is the characteristic coordinate of an arbitrary point of the apparatus cross-section, u_r is the local velocity of the fluid motion, and v_r is the local absolute velocity of particle motion.

Equilibrium of the particle in such conditions is expressed as

$$\frac{\pi d^3}{6} g(\rho - \rho_0) = \lambda \frac{\pi d^2}{4} \rho_0 \frac{(u_r - v_r)^2}{2} \tag{10.2}$$

or

$$u_r - v_r = w\sqrt{\frac{4}{3}B}, \tag{10.3}$$

where $B = \frac{\rho - \rho_0}{\rho_0} \frac{gd}{w^2}$ is a generalised parameter of separation curves in critical regimes, ρ is the density of the particle material, and w is the mean velocity of fluid flow in the channel.

It follows from Eq. (10.2) that

$$(u_r - v_r)^2 = \frac{4gd}{3\lambda} \frac{(\rho - \rho_0)}{\rho_0}. \tag{10.4}$$

For air, $\rho_0 \ll \rho$; $B = \frac{\rho}{\rho_0} \frac{gd}{w^2}$.

We examine the regime of turbulent flow around the particle characterised by the drag coefficient $\xi \cong 0.5$. In this case, the Reynolds criterion of the flow around a particle is

$$\mathrm{Re}_0 = \frac{(u_r - v_r) d \rho_0}{\mu} \geq 500. \tag{10.5}$$

To obtain an expression for B in the right-hand side of the dependence Eq. (10.4), we divide both parts by w^2 (with w being the mean velocity with respect to channel cross-section). Taking this into account,

$$u_r - v_r = w\sqrt{\frac{4}{3\lambda}B}. \tag{10.6}$$

In this case, Reynolds criterion for a particle in the case of turbulent flow around it is $\mathrm{Re}_0 \geq 500$. Taking Eq. (10.3) into account, we obtain

$$\frac{w\sqrt{\frac{4}{3\lambda}B}\cdot d\rho_0}{\mu} \geq 500. \tag{10.7}$$

Condition Eq. (10.7), taking into account $\lambda = 0.5$, corresponds to

$$\sqrt{\frac{8}{3}Ar} \geq 500,$$

where $Ar \geq 93{,}750$, and where the Archimedes criterion

$$Ar = \frac{gd^3\rho\rho_0}{\mu^2} \tag{10.8}$$

and where μ is the dynamic viscosity of the medium.

From Eq. (10.7), the size of particles characterised by turbulent flow around them is

$$d \geq \frac{500\mu}{w\sqrt{\frac{4}{3}B}\cdot\rho_0}[m]. \tag{10.9}$$

Thus, quartzite ($\rho = 2650$ kg/m^3) particles with the size $d \geq 1$ mm are in the turbulent flow regime around them. We obtain from condition Eq. (10.7)

$$\frac{\rho_0}{\mu} \geq \frac{500}{w\sqrt{\frac{4}{3}B}\cdot d}.$$

The Reynolds criterion for a moving medium in a channel is

$$\text{Re} = \frac{wD\rho_0}{\mu}, \tag{10.10}$$

where D is the channel diameter.

Substituting the previous expression into this one, we obtain

$$\text{Re} \geq \frac{D}{d}\frac{500}{\sqrt{\frac{4}{3}B}}. \tag{10.11}$$

Condition Eq. (10.11) allows one to evaluate the fluid motion regime in a channel at a turbulent flow around the particles. Obviously, $(\text{Re})_{\text{min}}$ should be evaluated with respect to d_{max}, for which the parameter $B = B_{\text{max}}$, too. It has been established experimentally that at the separation of coarse-grain materials, $B_{\text{max}} = 2.5 \div 3$ for all mono-fractions and depends insignificantly on the apparatus design. The ratio D/d is on the order of 10^2 for experimental apparatuses, and an order of magnitude greater for industrial ones. Taking into account that $\lambda = 0.5$, we obtain from Eq. (10.11)

$$(\text{Re})_{\text{min}} \geq 1.7\cdot 10^4.$$

This has been confirmed experimentally. Thus, for equilibrium apparatuses $D = 100$ mm, $H = 900$ mm, and $H = 600$ mm for periclase ($\rho_m = 3600$ kg/m^3) classification at a velocity of $w = 1.5$ m/s, the fine-product output was $\gamma_f < 5\%$. In addition, the consumed concentration was maintained at a level of 1.5 kg/m^3, whereas the granulometric composition was characterised by an 11% content of particles less than 150 mm. One can check easily that, in such conditions, the Reynolds criterion is $\mathrm{Re}_0 = 10^4$. Thus, in the conditions of turbulent flow around particles (i.e., during separation of relatively coarse material), the entire process (from $\gamma_f = 0$) occurs actually in a turbulent regime of medium motion in the apparatus (with γ_f being the fine-product output).

Now we examine flow behaviour in a laminar flow regime around particles, which can take place at the separation of finely dispersed material. In this case, the following is valid:

$$\mathrm{Re}_0 = \frac{(u_r - v_r)d\rho_0}{\mu} \le 1. \tag{10.12}$$

It follows from equilibrium conditions that

$$\frac{\pi d^3}{6} g\rho = 3\pi\mu(u_r - v_r)d.$$

Taking Eqs. (10.8) and (10.12) into account, this equation is transformed into

$$Ar = 18\mathrm{Re}.$$

According to Eq. (10.12), we can write

$$Ar \le 18.$$

In air, this is valid for particles of the size

$$d \le 10^{-3}\sqrt{\frac{0,4954}{\rho}}.$$

For example, for quartzite, this is valid for particles of the size $d \le 57$ μkm.

It follows from inequality Eq. (10.12) that

$$\frac{\rho_0}{\mu} \le \frac{1}{(u_r - v_r)d}. \tag{10.13}$$

We then substitute the obtained result into the expression of Re:

$$\mathrm{Re} \le \frac{D}{d}\frac{w}{(u_r - v_r)}. \tag{10.14}$$

To express the ratio $\frac{u_r - v_r}{w}$, we use relation Eq. (10.4), taking into account that the drag coefficient in a laminar flow around a particle is

$$\lambda = \frac{24}{\mathrm{Re}} = \frac{24\mu}{(u_r - v_r)d\rho_0}.$$

Then,

$$\frac{u_r - v_r}{w} = \sqrt{\frac{4}{3}B\frac{(u_r - v_r)d\rho_0}{24\mu}}.$$

Hence,

$$\frac{u_r - v_r}{w} = \sqrt{\frac{1}{18}\mathrm{Re}B}.$$

Taking into account that $\mathrm{Re}^2 B = Ar$, we obtain

$$\frac{u_r - v_r}{w} = \frac{\sqrt{ArB}}{18}. \tag{10.15}$$

Substituting the obtained expression Eq. (10.14) into Eq. (10.15), we obtain

$$\mathrm{Re} \le \frac{D}{d}\frac{18}{\sqrt{ArB}}.$$

Taking into account that, in the laminar regime $Ar \le 18$, we finally obtain

$$\mathrm{Re} \le \frac{D}{d}\sqrt{\frac{18}{B}}. \tag{10.16}$$

During displacement of finely dispersed material in the channel, the ratio $\frac{D}{d}$ is on the order of $10^3 \div 2 \cdot 10^3$, and it was established experimentally that $B_{\max}$ at $[F_f(x) = 0]$ has different values for various monofractions. Here, $F_f(x)$ is particle extraction into the fine product:

$$F_f(x) = \gamma_f \frac{r_f}{r_c},$$

where r_c is the narrow class content in the initial material, r_f is the content of the same class in the fine product, and γ_f is the fine-product output.

Thus, laminar flow around particles occurs even at a sufficiently developed flow turbulence overlapping the area of transient regimes.

We evaluate Re for individual particles characterised by the Archimedes criterion value $Ar = 18$. It has been established experimentally that

$$B_{\max}[F(x) = 0] = 2.0$$

and

$$B_{\min}[F(x) = 100\%] = 0.5.$$

Then, the ratio Eq. (10.16) acquires the values

$$\mathrm{Re} \cong 10^3 \div 10^4.$$

According to experimental data on powdered aluminium ($\rho = 2700\ \mathrm{kg/m^3}$) separation in a shelf apparatus with $z = 10$, $i = 5$, and $\mu_p = 1.5 \div 2.2\ \mathrm{kg/m^3}$at an airflow velocity of $w = 0.27$ m/s, the monofraction $d = 56.5$ μm ($A = 18$) had a fractional extraction $F_f(x_0) = 0$. At a velocity of $w = 1.7$ m/s, $F_f(x) = 100\%$.

In this case, the fluid flow regime in the apparatus is characterised numerically, respectively, by

$$\text{Re} = 1.2 \cdot 10^3 \text{ and } \text{Re} = 8 \cdot 10^3.$$

Thus, with regard to the classification of particle overflow in a laminar regime, the entire process is characterised by both laminar and transient, and up to turbulent regimes. In real conditions, a significant number of mono-fractions overflows in the intermediate regime ($18 < Ar < 93{,}750$), which should cause intermediate and turbulent regimes of the medium motion in the apparatus. To clarify this issue is extremely important for prognostic estimation of powder distribution in flows and obtained products.

All these conclusions refer to individual particles. Regularities for mass motion of particles can be somewhat different. It is important to clarify this, because affinization parameters of separation curves depend on it. By way of example, we can show the results of aluminium powder distribution in a flow. The content of particles less than 45 μm in the flow was 30%. In experiments, the size range of 0–45 μm was separated into nine narrow classes, and the entire size range of this powder was analysed as 20 narrow classes. The coarsest particles in this mixture were 400 μm. The experiments were performed at flow velocities equal to 1.46 m/s, 1.19 m/s, 0.92 m/s, 0.65 m/s, 0.53 m/s, 0.38 m/s, 0.31 m/s, and 0.29 m/s. Then, quartz powder, with a size from 100 μm–3 mm, was separated in the same channel. The fractional extraction dependency on the affinization parameter B for various size classes is shown in Figure 10.1. In this plot, the difference between regularities of separation in turbulent and laminar flow regimes and particle overflow is seen clearly. A reliable universal separation curve is obtained for coarse quartzite powder particles, and at elevated velocities of 1.19 m/s and 1.46 m/s in the case of aluminium powder separation.

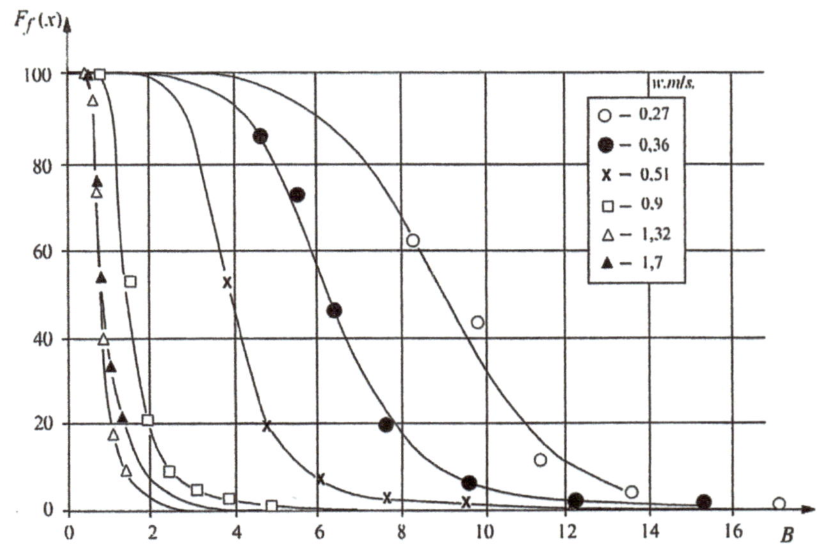

FIGURE 10.1 Dependency $F_f(x) = f(B)$ for aluminium powder.

However, affinization is violated for finer particles and low flow velocities, because the lower the flow velocity, the more intensely the curves diverge. Therefore, it becomes necessary to determine a generalising parameter for laminar flows, as well, side by side with the criterion B meant for turbulent flows, which generalises experimental data well in this regime.

2. Particle extraction in two-phase flow separation regimes

We can assume that distribution of particles of any narrow size class over the cross-section of a vertical channel is uniform, especially at a certain distance from the site of solid-phase entry into the flow.

As known from hydraulics, the distribution of local velocities of a continuous phase depends on the geometric characteristics of a channel. In a general form, this distribution can be expressed by the relationship (Figure 10.1)

$$w_r = w \cdot f\left(\frac{r}{R}\right), \tag{10.17}$$

where R is the channel diameter.

As usual, we take the total number of particles of the size class under study in a certain segment of the flow as unity. Then, fractional extraction from this segment equals the number of particles of this class with an absolute positive velocity that is more than or equal to zero.

We take the direction of ascending vertical flow as positive. According to Eq. (10.5), for a cross-section point with absolute velocity $v_r \geq 0$, we can write

$$u_r \geq w\sqrt{\frac{4B}{3\lambda}}. \tag{10.18}$$

Substituting Eq. (10.18) into Eq. (10.17), we obtain

$$f\left(\frac{r}{R}\right) \geq \sqrt{\frac{4}{3\lambda} \cdot B}. \tag{10.19}$$

Similarly, for particles with $v_r \leq 0$,

$$f\left(\frac{r}{R}\right) \leq \sqrt{\frac{4B}{3\lambda}}. \tag{10.20}$$

Inequalities Eqs. (10.19) and (10.20) comprise the following limiting cases:

$$f\left(\frac{r}{R}\right) > \sqrt{\frac{4B}{3\lambda}}.$$

In this case, we obtain the distribution coefficient $F_f(x) = 1$ for the corresponding coordinates. Respectively, at $f\left(\frac{r}{R}\right) < \sqrt{\frac{4B}{3\lambda}}$, we obtain $F_f(x) = 0$ for any coordinate r. In the intermediate case, certain coordinates form level lines according to the equality

$$f\left(\frac{r_0}{R}\right) = \sqrt{\frac{4B}{3\lambda}}. \tag{10.21}$$

Assume that this equation has one real root:

$$\frac{r_0}{R} = f^{-1}\left(\frac{4B}{3\lambda}\right). \tag{10.22}$$

Taking this into account, we find the corresponding area S_{r_0}, for which the following is valid:

$$f\left(\frac{r_i}{R}\right) \geq f\left(\frac{r_0}{R}\right).$$

In all this area, particles of size r_0 ascend, and those beyond it descend. Then, the distribution coefficient can be written as

$$F_f(x) = \frac{\omega_{r_0}}{\omega_R} = c\left(\frac{r_0}{R}\right) \text{ for a convex profile } f\left(\frac{r}{R}\right)$$

and

$$F_f(x) = C\left[1 - \left(\frac{r_0}{R}\right)^2\right] \text{ for a concave profile } f\left(\frac{r}{R}\right).$$

The coefficient C characterises the shape of level lines and the flow area—for instance, for a circle $c = 1$. Substituting Eq. (10.19) into the obtained equations, we finally obtain

$$F_f(x) = \varphi\left(\frac{B}{\lambda}\right). \tag{10.23}$$

A similar dependency is valid for two and more roots of Eq. (10.20), which takes place in case of complicated profiles $f\left(\frac{r}{R}\right)$.

A specific expression for the distribution coefficient can be obtained passing to the specific profile of a continuous medium. We examine, in turn, possible cases of particle and flow interaction. Let the regime of the medium motion represent a developed turbulence. In this case, the distribution of continuous medium velocities over the cross-section can be expressed by the dependency (10.21). According to Eq. (10.21), we can obtain

$$\frac{n+2}{n}\left[1 - \left(\frac{r_0}{R}\right)^n\right] = \sqrt{\frac{4B}{3\lambda}}. \tag{10.24}$$

Hence,

$$\left(\frac{r_0}{R}\right)^n = 1 - \frac{n}{n+2}\sqrt{\frac{4B}{3\lambda}}. \tag{10.25}$$

Here, the cross-sectional area for which $w\left(\frac{r}{R}\right) \geq w\left(\frac{r_0}{R}\right)$ is $S_0 = \pi r_0^2$.

Respectively, fractional extraction is

$$F_f(x) = \left(\frac{r_0}{R}\right)^2.$$

Taking Eq. (10.25) into account, we obtain

$$F_f(x) = \left[1 - \frac{n+2}{n}\sqrt{\frac{4B}{3\lambda}}\right]^{\frac{2}{n}}. \tag{10.26}$$

This dependency for fractional extraction in the form $F_f(x) = \varphi\left(\frac{B}{\lambda}\right)$ is characteristic of turbulent flow around the particles.

A laminar flow regime around all narrow classes is characterised by a drag coefficient equal to $\lambda = \frac{24}{\mathrm{Re}_{w_0}}$.

As shown, in this case the regime of the medium motion extends from laminar to turbulent regions. Recall that dependency (10.22) characterises all regimes of medium flow. Substituting the drag coefficient value into Eq. (10.24), we obtain

$$\frac{n+2}{n}\left[1 - \left(\frac{r_0}{R}\right)^n\right] = \sqrt{\frac{4B}{3}\frac{w_{r_0} d\rho_0}{24\mu}}.$$

Using this dependency, we can obtain, in general form,

$$\frac{n+2}{n}\left[1 - \left(\frac{r_0}{R}\right)^n\right] = \sqrt{\frac{\frac{n+2}{n} w\left[1 - \left(\frac{r_0}{R}\right)^n\right] d\rho_0}{18\mu} B}. \tag{10.27}$$

The radicand can be simplified by squaring both parts. We finally obtain

$$\frac{n+2}{n}\left[1 - \left(\frac{r_0}{R}\right)^n\right] = \frac{\mathrm{Re}_{w_0} \cdot B}{18}.$$

Taking into account that $\mathrm{Re}_{w_0}^2 \cdot B = Ar$, we arrive at

$$\frac{n+2}{n}\left[1 - \left(\frac{r_0}{R}\right)^n\right] = \frac{\sqrt{Ar \cdot B}}{18}, \tag{10.28}$$

and because

$$F_f(x) = \left(\frac{r_0}{R}\right)^2,$$

we finally obtain

$$F_f(x) = \left[1 - \frac{\sqrt{Ar \cdot B}}{18} \cdot \frac{n}{n+2}\right]^{\frac{2}{n}}. \tag{10.29}$$

For a parabolic profile, $n = 2$, and the dependency Eq. (10.29) is transformed into

$$F_f(x) = 1 - \frac{\sqrt{Ar \cdot B}}{36}. \tag{10.30}$$

As already shown, in real conditions, the flow around a significant part of narrow classes is in the transient regime. The dependency of drag coefficients on Re and Ar criteria is known:

$$\lambda = \frac{4}{3}\frac{Ar}{\mathrm{Re}^2}. \tag{10.31}$$

There is an interpolation dependency of the following form:

$$\mathrm{Re} = \frac{Ar}{18 + 0.61\sqrt{Ar}}. \tag{10.32}$$

We substitute Eq. (10.33) into Eq. (10.31)

$$\lambda = \frac{4}{3}\frac{\left(18 + 0,61\sqrt{Ar}\right)^2}{Ar},$$

and substituting this into Eq. (10.24), we obtain

$$\frac{n+2}{n}\left[1 - \left(\frac{r_0}{R}\right)^n\right] = \sqrt{\frac{Ar \cdot B}{\left(18 + 0,61\sqrt{Ar}\right)^2}}.$$

Now we pass to fractional extraction:

$$\frac{n+2}{n}\left(1 - F_f(x)^{\frac{n}{2}}\right) = \frac{\sqrt{Ar \cdot B}}{18 + 0.61\sqrt{Ar}},$$

where we obtain a generalised expression for fractional extraction in an arbitrary regime of flow around the particles:

$$F_f(x) = \left[1 - \frac{n}{(n+2)}\frac{\sqrt{Ar \cdot B}}{\left(18 + 0.61\sqrt{Ar}\right)}\right]^{\frac{2}{n}}. \tag{10.33}$$

As a result of numerous studies, it has been established that in turbulent and transient two-phase flow regimes, the flow profile is somewhat extended in its central part and acquires a shape close to a parabolic profile as a result of the presence of the solid phase in it. Under certain assumptions, we can believe that all three two-phase flow regimes are characterised by $n = 2$. Then, for an arbitrary overflow regime,

$$F_f(x) = \left(1 - \frac{\sqrt{Ar \cdot B}}{36 + 1.22\sqrt{Ar}}\right). \tag{10.34}$$

In turbulent regimes, $Ar \geq 10^5$; therefore, the quantity 36 in the denominator can be neglected, and then for this regime

$$F_f(x) = \left(1 - \sqrt{0.4B}\right). \tag{10.35}$$

Here, fractional extraction is connected unambiguously with the affinization parameter determined through the two-phase flow structure, which appears here as an invariant of solid-phase distribution. Fractional extraction of a coarse product is easier to visualise, because

$$F_f(x) + F_c(x) = 1,$$

and therefore

$$F_c(x) = \sqrt{0.4B}. \tag{10.36}$$

From Eqs (10.35), (10.36), and relation Eq. (10.34), we can formulate the affinization criteria for all two-phase flows regimes:

1. or a turbulent flow:

$$B_1 = 1 - F_f(x) = \sqrt{0.4B}.$$

   Here it follows *per se* that B_1 is proportional to the empirically found parameter B.

2. or a laminar flow:

$$B_2 = \frac{\sqrt{Ar \cdot B}}{36}. \tag{10.37}$$

3. or an arbitrary flow:

$$B_3 = \frac{\sqrt{Ar \cdot B}}{36 + 1.22\sqrt{Ar}}. \tag{10.38}$$

Thus, it is confirmed again that the empirically found parameter

$$B = \frac{gx}{w^2}\frac{(\rho - \rho_0)}{\rho_0}$$

is an affinization parameter for a turbulent flow regime around particles.

For a laminar flow regime, taking into account the fact that

$$Ar = \frac{gx^3(\rho - \rho_0)\rho_0}{\mu^2},$$

the affinization parameter acquires the form

$$B_2 = \frac{gx^2(\rho - \rho_0)}{\mu \cdot w}. \tag{10.39}$$

As already noted, the entire powder distribution range can be covered by two parameters characterising turbulent and laminar flow regimes.

It is of interest to analyse the physics of dependency Eq. (10.39). For this purpose, we divide

$$\frac{B_2}{B} = \frac{gx^2(\rho - \rho_0)w^2\rho_0}{\mu \cdot wgx(\rho - \rho_0)} = \frac{dw\rho_0}{\mu} = \mathrm{Re}$$

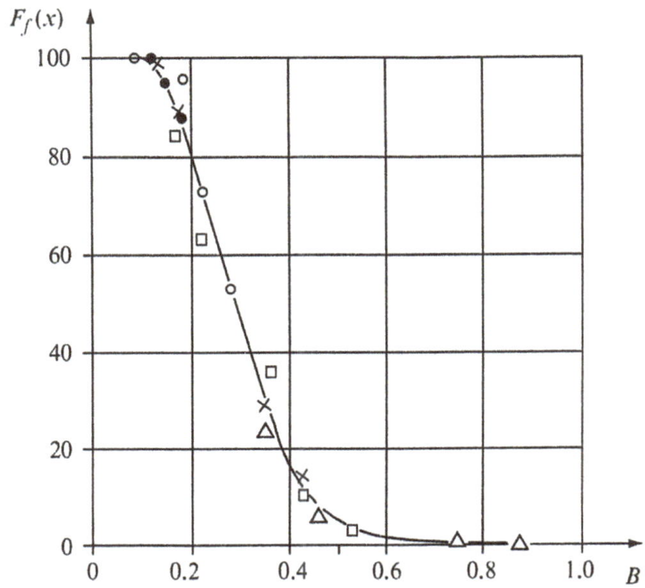

FIGURE 10.2 Fractional separation of chromium oxide as a function of the parameter *B*.

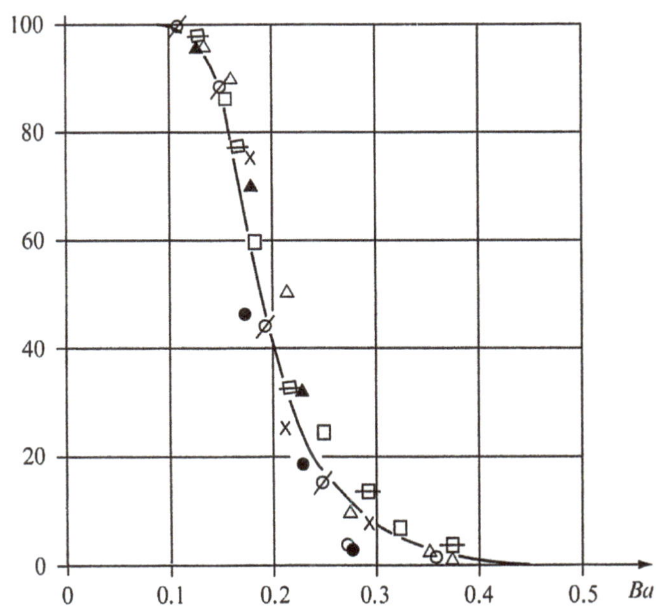

FIGURE 10.3 Fractional separation of aluminium powder as a function of the parameter B_α.

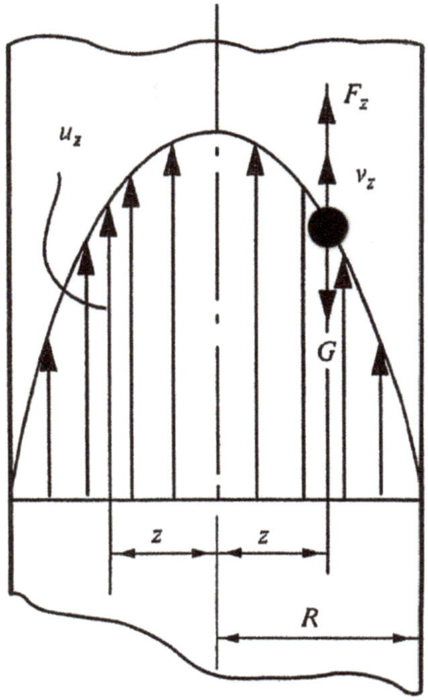

FIGURE 10.4 Determining the distribution coefficient.

and obtain the Reynolds criterion for a particle. Hence, we can write

$$B_2 = B \cdot \text{Re}. \tag{10.40}$$

Relation Eq. (10.40) demonstrates the physical meaning of parameter B_2.

In turbulent flow regimes around particles, the Re criterion degenerates and the drag coefficient λ acquires a constant value. Here, the dynamics of the process is described fully by the parameter B.

To obtain universal curves for the laminar flow region, where particle resistance is inversely proportional to Re, it is necessary to take into account both parameters (i.e., their product, $B \cdot \text{Re}$).

As experiments have demonstrated, these two criteria overlap the entire range of two-phase flows.

To illustrate the use of these criteria, Figure 10.2 shows the affinization of separation results of milled chromium oxide ($\rho = 3600$ kg/m^3) with a particle size of 0.025–3.75 mm using the criterion B. Figure 10.3 shows the results of affinization for separation curves for aluminium powder ($\rho = 2700$ kg/m^3) using the criterion B_α As Figure 10.4 shows, the criterion B does not provide affinization of these curves.

CHAPTER

Basic Physics of Cascade Separation

11

CHAPTER OUTLINE

1. Main principles of cascade mass-exchange processes 175
2. Discrete model 180
3. Analysis of a mathematical model of a regular cascade 187
4. Separation at cyclic feed of initial material into cascade apparatus 191
5. Absorbing Markov chain in cascade classification of bulk materials 193
6. Mathematical model of cascade separation with arbitrary distribution coefficients 195
7. Discrete model of critical regimes of vertical two-phase flows 200
8. Optimisation of principal parameters of multi-stage separation 212

1. Main principles of cascade mass-exchange processes

In recent years, it has been established as a result of intense experimental studies that multistage organisation of bulk materials separation allows a significant increase in the process efficiency. This principle has proved to be much more efficient than the existing methods of one-stage separation. It ensures a considerable increase of productivity and quality of products produced in apparatuses of moderate sizes. For instance, we have fabricated a cascade classifier (0.6 m × 1.1 m × 5 m) with the output 140 t/hour and separation boundary of 100 μm for potassium plant at the Dead Sea Works operating since 2002.

Many technological processes, such as petroleum cracking, gas adsorption, isotopes or liquid mixtures separation, are organised on cascade principle. Therefore, development and solution of mathematical models and search for invariants for cascade separation are basically of phenomenological character. The obtained results of powder separation can be referred to a group of cascade processes of a different physical nature.

A proportional model reflecting mass exchange between stages (Figure 11.1) is the best illustration of materials separation on the basis of cascade principle.

The value characterising narrow size class separation in a separate cascade unit can be represented in the simplified form as:

$$K = \frac{r_i^*}{r_i}$$

Entropic Invariants of Two-Phase Flows. http://dx.doi.org/10.1016/B978-0-12-801458-5.00011-3

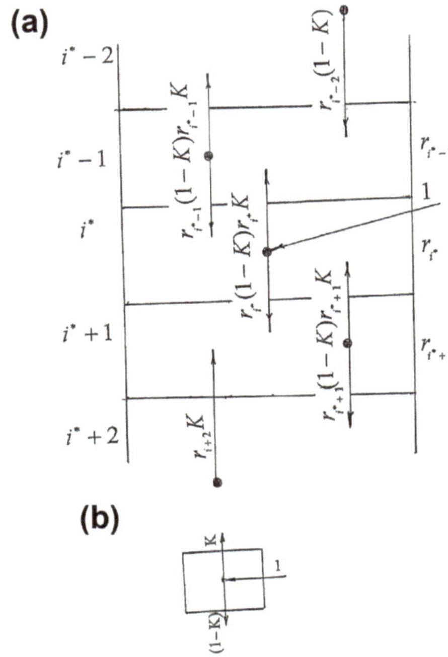

FIGURE 11.1 Schematic diagram of narrow-class particles distribution: (a) in a cascade; (b) in a stage.

where r_i is the initial content of narrow size class particles on the i-th stage; r_i^* – the number of particles of the same size class passing from the i-th to the (i – 1)-th stage; and K– distribution coefficient.

At the same structure of stages for each size class, the distribution coefficient is constant and independent of the stage number. This was confirmed in special experiments on a cascade separator comprising 16 stages.

A proportional model of distribution of particles of a certain fixed size class over the apparatus height with feeding to the i^*-th stage is shown in Figure 11.1(a).

The character of classification is determined, in many respects, by the place of material feed into the flow. Let the initial mixture contain a certain number of particles of i-th size. We take their initial content as unity. Fractional separation of a fixed narrow size class depending on the number of stages is determined, at invariable parameters of classifier operation regime, by the number of stages.

We call the fraction $F_f(x) = r_f/r_s$ fractional extraction of fine product for a narrow size class; here r_f and r_s are the amounts of narrow size class in the fine product and initial material, respectively.

In case of one stage, the process pattern is rather simple (see Figure 11.1(b)). Here, fractional extraction of fine product corresponds to the distribution coefficient $F_f(x)_1 = K$.

For two separation stages, the distribution pattern is as shown in Figure 11.2. Bulk material is fed to the first stage. For computation convenience, the amount of particles of a narrow size class is taken as unity. Its actual content is taken into account in final computation. It is clear from the scheme that, at the first stage, the fraction K ascends, while the fraction (1 – K) descends. At the second stage, the

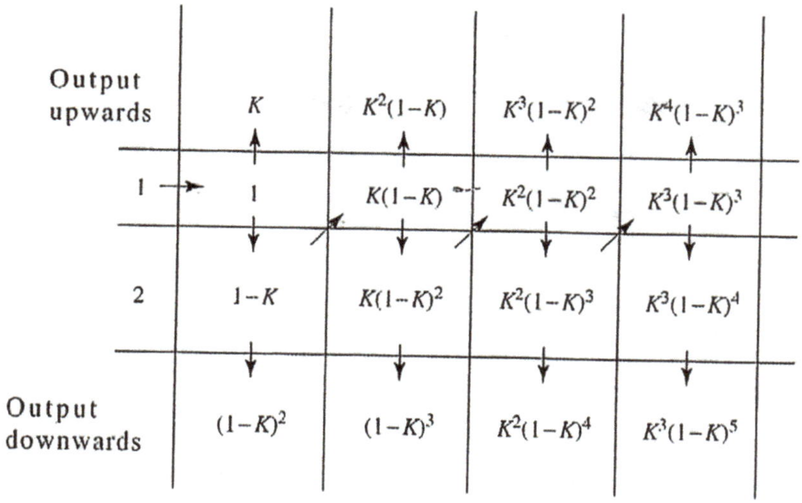

FIGURE 11.2 Schematic diagram of narrow-class distribution in two separation stages.

fraction $K(1-K)$ ascends, while the fraction $(1-K)^2$ descends. Then the distribution goes on according to this scheme.

Fractional extraction for two stages can be represented as a sum of an infinite series:

$$F_f(x)_2 = \lim_{n\to\infty}\left[K + K^2(1-K) + K^3(1-K)^2 + \ldots + K^n(1-K)^{n-1}\right]. \tag{11.1}$$

In a generalised form:

$$F_f(x)_2 = \lim \sum_{n=1}^{\infty}(1-K)^{n-1}K^n. \tag{11.2}$$

For three separation stages, the distribution pattern acquires the form shown in Figure 11.3. Consequently:

$$F_f(x)_3 = \lim_{n\to\infty}\left[K + K^2(1-K) + 2K^3(1-K)^2 + \ldots + \underbrace{2^{n-1}(1-K)^nK^{n+1}}_{A_n}\right]. \tag{11.3}$$

We find $F_f(x)_2$ and $F_f(x)_3$:

$$F_f(x)_2 = \lim_{n\to\infty}\left[K + K^2(1-K) + K^3(1-K)^2 + \ldots + K^n \times (1-K)^{n-1}\right]$$

$$= \lim_{n\to\infty}\left[K + K^2(1-K) + K^3(1-K)^2 + \ldots + K^{n+1}(1-K)^n\right],$$

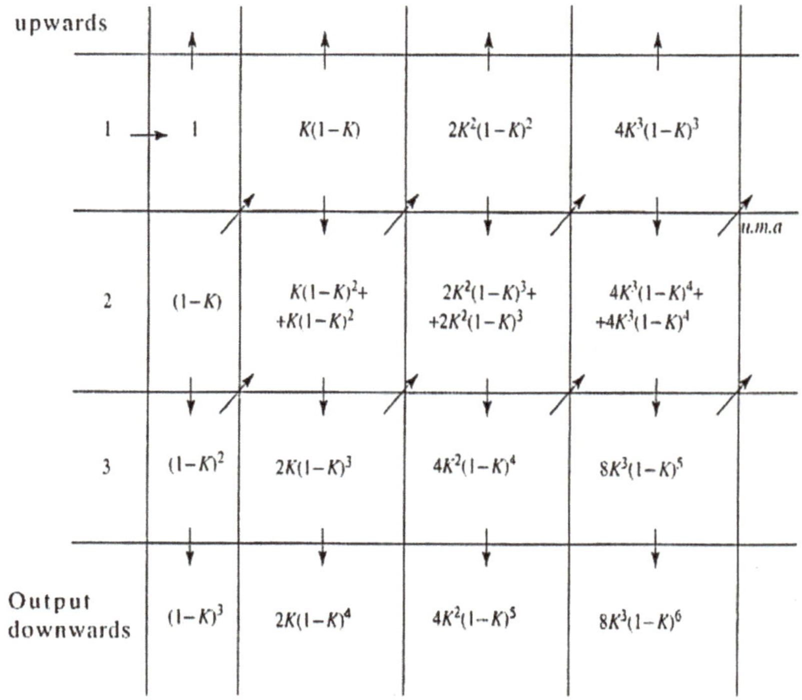

FIGURE 11.3 Schematic diagram of narrow-class distribution in three separation stages.

since

$$\lim_{n\to\infty} K^{n+1}(1-K)^n = 0.$$

We sum up these two expressions aslant:

$$K + K\left(1+K-K^2\right) + K^2(1-K)\left(1+K-K^2\right) + K^3(1-K)^2 \\ \times\left(1+K-K^2\right) + \ldots + K^n(1-K)^{n-1}\left(1+K-K^2\right) = 2F_f(x)_2,$$

factor the common multiplier out of the left-hand side:

$$K + \left(1+K-K^2\right) \times \underbrace{\left[K + K^2(1-K) + K^3(1-K)^2 + \ldots + K^n(1-K)^{n-1}\right]}_{F_f(x)_2} = 2F_f(x)_2$$

and, hence, obtain:

$$K + \left(1+K-K^2\right)F_f(x)_2 = 2F_f(x)_2$$

or

$$F_f(x)_2 = \frac{K}{1+K^2-K}.$$

We transform the obtained expression:

$$\frac{K}{1+K^2-K} = \frac{K}{1-K(1-K)} = \frac{K}{1-F_f(x)_1(1-K)} = 2F_f(x)_2.$$

Similarly, we find the dependence for fractional extraction in case of three separation stages:

$$\lim_{n\to\infty}\left[K + K^2(1-K) + 2K^3(1-K)^2 + 4K^4(1-K)^3 + \ldots + 2^{n-1}K^{n+1}(1-K)^n\right] = F_f(x)_3$$

$$\lim_{n\to\infty}\left[K + K^2(1-K) + 2K^3(1-K)^2 + 4K^4(1-K)^3 + \ldots + 2^nK^{n+2}(1-K)^{n+1}\right] = F_f(x)_3$$

Summing up these two expressions, we obtain:

$$K - K^2(1-K) + K\left(1+2K-2K^2\right) + K^2(1-K)\left(1+2K-2K^2\right) + 2K^3(1-K)^2$$
$$\times\left(1+2K-2K^2\right) + \ldots + 2^{n-1}K^{n+1}(1-K)^n\left(1+2K-2K^2\right) = 2F_f(x)_3$$

or

$$K - K^2(1-K) + \left(1+2K-2K^2\right)F_f(x)_3 = 2F_f(x)_3;$$

$$F_f(x)_3 = \frac{K-K^2(1-K)}{1+2K^2-2K} = K\frac{1+K^2-K}{1+2K^2-2K} \tag{11.4}$$

We transform the expression for fractional extraction in case of three stages:

$$K\frac{1+K^2-K}{1+2K^2-2K} = \frac{K}{\frac{1+2K^2-2K}{1+K^2-2K}} = \frac{K}{\frac{1+K^2-K+K^2}{1+K^2-K}}$$
$$= \frac{K}{1-\frac{K(1-K)}{1+K^2-K}} = \frac{K}{1-F_f(x)_2(1-K)}$$

We can assume that, in a general case, fractional extraction for an apparatus of n stages is written as:

$$F_f(x)_n = K\frac{1}{1-F_f(x)_{n-1}(1-K)}. \tag{11.5}$$

Material distribution over the apparatus height for four separation stages leads to an infinite convergent series, whose every term represents an odd number from the Fibonacci series. In this case, fractional extraction has a form that is difficult to express by a final result:

$$F_f(x)_4 = K + \sum_{n=1}^{\infty} A^f_{2n-1}K^{n+1}(1-K)$$

Similarly, fractional extractions for five, six, seven and any other finite number of the classifier stages are also difficult to express in a finite form. For example:

$$F_f(x)_5 = K + K^2(1-K) + \sum_{n=1}^{\infty} A_n K^{n+2}(1-K)^{n+1},$$

$$\text{where} \quad A_n = A_{n+1} + 3^{n-1}; \ A_0 = 2;$$

$$F_f(x)_6 = K + K^2(1-K) + 2K^3(1-K)^2 + 5K^4(1-K)^3 + 14K^5(1-K)^4 + 42K^6(1-K)^5 + 131K^7(1-K)^6 + 420K^8(1-K)^7 + \ldots;$$

$$F_f(x)_7 = K + K^2(1-K) + 2K^3(1-K) + 5K^4(1-K)^3 + 14K^5(1-K)^4 + 42K^6(1-K)^5 + 132K^7(1-K)^6 + 428K^8(1-K) + \ldots;$$

$$F_f(x)_8 = K + K^2(1-K) + 2K^3(1-K) + 5K^4(1-K)^3 + 14K^5(1-K)^4 + 42K^6(1-K)^5 + 132K^7(1-K)^6 + 429K^8(1-K)^7 + \ldots$$

In a general case, it is not always possible to express the n-th term of these series, not to speak about the transition to a sum. However, such formula was found for a cascade of eight stages in the following form:

$$A_{n(8)} = \frac{2n!}{n!(n+1)!}$$

Obviously, although the proportional model illustrates well the mechanism of the process, it is very cumbersome and inconvenient for computations. Therefore, we examine another approach to this problem.

2. Discrete model

We make an attempt to prove main dependencies for a separating cascade model in a different way. For this purpose, we consider cascade distribution of material at discrete moments of time (acts) with equal intervals. We assume that the material is fed to the apparatus in equal batches with a certain time interval. As known, at a definite solid phase concentration $\mu \leq 2$ kg/m^3, any size class is separated as if other particles do not take part in the process.

On the basis of the assumptions above, we prove the existence of $\lim_{n\to\infty} r_{ij}^m$ (i.e. that in the course of separation process, the material quantity gets stabilised at each stage). Here, m is the number of material distribution act, r_{ij}^m – the quantity of material of size class j at the i-th stage during the distribution act m.

Let $r_j^m = (r_{1j}^m, r_{2j}^m, r_{3j}^m, \ldots, r_{zj}^m)$ by definition, mass exchange balance $r_{ij}^{m+1} = K_j r_{i^*+1,j}^m + (1-K_j) r_{i^*-1,j}^m$ holds true in adjacent stages for $i \neq 1, z, i^*$ (i^* being the place of material feed into the apparatus).

Assume that the material is fed into the apparatus in equal batches. Let the quantity of one batch be γ. Then the relation:

$$r_{i^*j}^{m+1} = K_k r_{i=1,j}^{m} + (1 - K_j) r_{i^*-1,j}^{m} + \gamma$$

holds for the i^*-th stage. For the first stage:

$$r_{1,j}^{m+1} = K_j r_{2,j}^{m}$$

and for the z-th stage:

$$r_{z,j}^{m+1} = (1 - K_j) r_{z-1,j}^{m}$$

Thus, we obtain the following iterative matrix relationship:

$$r_j^{m+1} = A r_j^m + b$$

$$A = \begin{vmatrix} 0 & K_j & 0 & \cdots & 0 & 0 \\ 1-K_j & 0 & K_j & \cdots & 0 & 0 \\ 0 & 1-K_j & 0 & K_j & \cdots & 0 \\ 0 & 0 & \cdot & \cdot & \cdot & \cdot \\ \cdot & \cdot & \cdot & \cdot & \cdot & \cdot \\ 0 & 0 & 0 & 0 & 1-K_j & 0 \end{vmatrix} \quad b = \begin{vmatrix} 0 \\ 0 \\ \cdot \\ \gamma \\ \cdot \\ 0 \end{vmatrix} \overset{i^*}{\leftarrow}$$

As known, $r_j^m \rightarrow r_j$ (where r_j is a solution of the iterative equation $r_j = A r_j + b$) if and only if $|\lambda| < 1$ is valid for all eigenvalues of A matrix. According to the Flobenius theorem, if a matrix consists of non-negative real elements and cannot be reduced to a block diagonal form by simultaneous transmutations, then $s \leq |\lambda| \leq S$ is valid for its modulo maximal eigenvalue, where $S = \max_i \sum a_{ij}, \quad s = \min_i \sum a_{ij}$, and if $s \neq S$, then $s < |\lambda| < S$.

For the matrix A:$S = 1$ $s = \min(K_j, 1 - K_j)$ (i.e. for any eigenvalue of A matrix, $|\lambda| < 1$ holds). Then $r_j^m \rightarrow r_j$, where r_j is the solution of equation $r_j = A r_j + b$.

Consider the process in a steady separation regime, r_i is the quantity of the size class j in the separation stage i in a steady state (we omit the index j for convenience). The obtained conclusion is true for any size class available at a given stage.

We derive a design formula for fractional extraction of size class j into the fine product. It follows from the definition of $F_f(x)$ that:

$$F_f(x) = \frac{r_f}{r_s} = \lim_{t \to \infty} \frac{\sum_{l=1}^{t} r_{f_1}^{t}}{\sum_{l=1}^{t} r_{s_1}}$$

where $r_{f_1}^{l}$ and r_{s_1} are quantities of size class j material in one batch leaving the apparatus (from the first stage upward) and entering the apparatus (into stage i*), respectively, during the separation act l. Here, we have established that r_{s1} is constant and independent of l. Let us prove that:

$$F_f(x) = \frac{r_f}{r_s} = \lim_{l \to \infty} \frac{r_{f_1}^{l}}{r_{s_1}}$$

It follows, as stated above, that $\lim_{l\to\infty}\frac{r_{f_1}^l}{r_{s_1}}$ exists (since $r_{f_1}^l = Kr_1^l$), and we call it α, consequently, $r_{f_1}^l = r_{s_1}\alpha + o(l)(l\to\infty)$. We substitute it into the equality $\frac{r_f}{r_s} = \lim_{t\to\infty}\frac{\sum_{l=1}^{t} r_{f_1}^l}{\sum_{l=1}^{t} r_{s_1}}$ and obtain $\frac{r_f}{r_s} = \lim_{l\to\infty}\frac{r_{f_1}^l}{r_{s_1}} = \alpha$. Therefore, $F_f(x) = \frac{r_f}{r_s} = \lim_{l\to\infty}\frac{r_{f_1}^l}{r_{s_1}}$ is valid for the size class j.

If K is the distribution coefficient of the size class j, the quantity of this size class rising from stage i to stage $i-1$ equals $K\cdot r_i$, while the quantity descending to stage $i+1$ equals $(1-K)\cdot r_i$.

On the basis of Figure 12.1(a), we normalise the quantity of size class j at each stage for each portion of material entering the apparatus (i.e. $R_t = \frac{r_i}{r_{s_1}}$). We call R_i a quantity of size class j in a stationary state normalised with respect to the quantity of this class in each portion of material fed to the apparatus. Consequently, $R_{i-1} = r_i/r_{s1}\cdot K = R_t\cdot K$, then the normalised quantity coming out into the fine product equals R_1K. On the other hand, it equals $\frac{r_{f_1}}{r_{s_1}}$ ($r_{f_1} = \lim_{l\to\infty} r_{f_1}^l$) (i.e. $R_1K = \frac{r_{f_1}}{r_{s_1}}$), then $F_f(x) = R_1K$. After finding a design formula for $F_f(x)$, it will be possible to find the quantity of size class j coming out into the fine product using equation $r_f = F_f(x)r_s$ (where r_s is initially known).

We prove that the quantity of fractional extraction into the fine product for the size class j can be calculated by the formula:

$$F_f(x) = \begin{cases} \dfrac{1-\sigma^{z+1-i^*}}{1-\sigma^{z+1}} & K \neq 0.5 \\ \dfrac{z+1-i^*}{z+1} & K = 0.5 \\ 0 & K = 0 \end{cases}$$

where $\sigma = (1-K)/K, i^*$ is the number of stage of the material feed into the apparatus. To find F_f, we find R_1 first. It has been already proved that the quantity of material of each size class at each stage of the apparatus remains constant during the process (i.e. the quantity of material entering and coming out of any stage of the apparatus during one redistribution act is stationary). We can write the following balance equation:

$$\underbrace{(1-K)R_i + KR_i}_{\text{Normalized quantity coming out of stage } i} = \underbrace{(1-K)R_{i-1} + KR_{i+1}}_{\text{Normalized quantity entering stage } i} \tag{11.6}$$

$i \neq 1, z$, because the material goes out of these stages, and $i \neq i^* - 1, i^*, i^* + 1$, since the material entering stage i^* directly influences these stages. All these cases are considered below.

Fractional extraction into the coarse product equals r_c/r_s, where r_c is the quantity of size class j in the coarse product output. Therefore, $r_f/r_s + r_c/r_s = r_f + r_c/r_s = 1$ and, hence, the degree of fractional extraction of size class j into coarse product equals $1 - F_f(x)$. It was proved earlier that $F_f(x) = KR_1$, similarly, we can obtain that $1 - F_f(x) = (1-K)R_z$, and then:

$$KR_1 + (1-K)R_z = F_f(x) + \left(1 - F_f\right) = 1 \tag{11.7}$$

Taking into account that $\sigma=(1-K)/K$, one can write (Eq. 11.7) as follows:

$$R_1+\sigma R_2=1+\sigma. \tag{11.8}$$

The quantity of each size class at each separation stage is constant during the entire process, and, hence, the following relations are valid:

- for the second stage:

$$R_2=R_1+\Delta R_2$$

- for the third stage:

$$R_3=R_2+\Delta R_3=R_1+\Delta R_2+\Delta R_3\ldots$$

- for the (i^*-1)-th stage:

$$R_{i^*-1}=R_{i^*-2}+\Delta R_{i^*-1}=R_1+R_2+R_3+\ldots+\Delta R_{i^*-1}$$

The quantity of material of size class j entering the apparatus equals $\frac{r_{s_1}}{r_{s_1}}=1$. Hence, for the i^*-th stage, the following is valid:

$$R_{i^*}=R_{i^*-1}+\Delta R_{i^*}+1=R_1+\Delta R_2+\Delta R_3+\ldots+\Delta R_{i^*}+1$$

for the (i^*+1)-th stage:

$$R_{i^*+1}=R_1+\Delta R_2+\Delta R_3+\ldots+\Delta R_{i^*+1}\ldots$$

for the z-th stage:

$$R_z=R_1+\Delta R_2+\Delta R_3+\ldots+\Delta R_z.$$

For the first stage, the following relation is valid:

$$R_1=R_2K=(R_1+\Delta R_2)K.$$

Hence,

$$R_1=\frac{1}{\sigma}\Delta R_2. \tag{11.9}$$

We can write for stage z:

$$R_z=(1-K)R_{z-1} \quad \text{and} \quad R_z=R_{z-1}+\Delta R_z.$$

From these two equalities, we obtain $R_{z-1}=(-1/K)\Delta R_z$ and substitute it into:

$$R_z=(1-K)R_{z-1}$$

Obtaining:

$$R_z = -\sigma \Delta R_z. \tag{11.10}$$

We consider a balance equation for stage $i \neq 1; z; i^* - 1; i^*; i^* + 1$:

$$(1-K)R_i + KR_i = (1-K)R_{i-1} + KR_{i+1}.$$

Substituting these two equalities into the balance Eq. (11.6), we obtain:

$$(1-K)(R_{i-1} + \Delta R_i) + K(R_{i-1} + \Delta R_i) = (1-K)R_{i-1} + K(R_{i-1} + \Delta R_i + \Delta R_{i+1}).$$

As known, $R_i = R_{i-1} + \Delta R_i$ and $R_{i+1} = \Delta R_i + \Delta R_{i+1}$,whence we obtain:

$$\Delta R_i(1-K) = K\Delta R_{i+1} \tag{11.11}$$

and then:

$$\Delta R = \frac{1}{\sigma}\Delta R_{i+1}. \tag{11.12}$$

We have already derived similar relations Eqs (11.9) and (11.10) for stages number 1 and z. Now we examine separation stage $i^* - 1$, for which we can write:

$$(1-k)R_{i^*-1} + KR_{i^*-1} = (1-K)R_{i^*-2} + KR_{i^*} + K$$

K is added here, because the normalised quantity of class j material entering the apparatus equals 1 and, hence, its part passing to stage $i^* - 1$ equals $1 \cdot K = K$. Therefore, by analogy with Eq. (11.11) we obtain:

$$\Delta R_{i^*-1}(1-K) = K\Delta R_{i^*} + K$$

and then:

$$\Delta R_{i^*-1} = \frac{1}{\sigma}(\Delta R_{i^*} + 1). \tag{11.13}$$

Considering stage $i^* + 1$:

$$(1-K)R_{i^*+1} + KR_{i^*+1} = (1-K)R_{i^*} + KR_{i^*+2} + (1-K)$$

we add $(1-K)$ because of the same reason as that for adding K to stage $i^* - 1$. By analogy with Eq. (11.11), we obtain:

$$\Delta R_{i^*+1}(1-K) = K\Delta R_{i^*+2} + (1-K)$$

and hence:

$$\Delta R_{i^*+1} = \frac{1}{\sigma}\Delta R_{i^*+2} + 1 \tag{11.14}$$

Balance equation for stage i^* is the same as for an ordinary stage i.

Similarly, we can write for stages $i^* - 1$ and $i^* + 1$:

$$(1 - K)R_{i^*} + KR_{i^*} = (1 - K)R_{i^*-1} + KR_{i^*+1}$$

whence follows:

$$\Delta R_{i^*} = \frac{1}{\sigma}\Delta R_{i^*+1}$$

Thus, we can obtain the following sequence of relations:

$$R_1 = \frac{1}{\sigma}\Delta R_2$$

$$\Delta R_2 = \frac{1}{\sigma}\Delta R_3$$

$$\Delta R_3 = \frac{1}{\sigma}\Delta R_4$$

$$\ldots$$

$$\Delta R_{i^*} - 2 = \frac{1}{\sigma}\Delta R_{i^*} - 1$$

$$\Delta R_{i^*} - 1 = \frac{1}{\sigma}(\Delta R_{i^*} + 1)$$

$$\Delta R_{i^*} = \frac{1}{\sigma}\Delta R_{i^*} + 1$$

$$\Delta R_{i^*} + 1 = \frac{1}{\sigma}R_{i^*} + 2 + 1$$

$$\Delta R_{i^*} + 2 = \frac{1}{\sigma}\Delta R_{i^*} + 1$$

$$\ldots$$

$$\Delta R_{z-1} = \frac{1}{\sigma}\Delta R_z$$

We substitute the second equation into the first:

$$R_1 = \frac{1}{\sigma^2}\Delta R_3$$

and the third into the obtained equation:

$$R_1 = \frac{1}{\sigma^3}\Delta R_4$$

and continue substituting in the same way:

$$\dots$$

$$R_1 = \frac{1}{\sigma^{i-1}} \Delta R_i$$

$$\dots$$

$$R_1 = \frac{1}{\sigma^{i^*} - 2} \Delta R_{i^*} - 1$$

$$R_1 = \frac{1}{\sigma^{i^*} - 1} (\Delta R_{i^*} + 1)$$

$$R_1 = \frac{1}{\sigma^{i^*}} \Delta R_{i^*} + 1 + \frac{1}{\sigma^{i^*} - 1}$$

$$R_1 = \frac{1}{\sigma^{i^*} + 1} \Delta R_{i^*} + 2 + \frac{1 + \sigma}{\sigma^{i^*}}$$

$$\dots$$

$$R_1 = \frac{1}{\sigma^{z-1}} \Delta R_z + \frac{1 + z}{\sigma^{i^*}} \tag{11.15}$$

From Eq. (11.10), $\Delta R_z = (-1/\sigma)R_z$, Substituting this into Eq. (11.15), we obtain:

$$R_1 = \frac{1}{\sigma^{z-1}} \cdot \Delta R_z + \frac{1 + \sigma}{\sigma^{i^*}}$$

It follows from Eq. (11.8) that $R_1 = 1 + \sigma - \sigma R_z$. Then we obtain the following system of equations:

$$\left. \begin{aligned} R &= -\frac{1}{\sigma^z} R_z + \frac{1 + \sigma}{\sigma^{i^*}} \\ R_1 &= 1 + \sigma - \sigma R_z \end{aligned} \right\} \tag{11.16}$$

and find R_1 from this system:

$$-\frac{1}{\sigma^z} R_z + \frac{1 + \sigma}{\sigma^{i^*}} = 1 + \sigma - \sigma R_z$$

$$\frac{1 + \sigma}{\sigma^{i^*}} - (1 + \sigma) = \frac{1}{\sigma^z} R_z - \sigma R_z$$

$$(1 + \sigma)\left(\frac{1 - \sigma^{i^*}}{\sigma^{i^*}}\right) = R_z\left(\frac{1 - \sigma^{z+1}}{\sigma^z}\right)$$

Hence:

$$R_z = \frac{(1 + \sigma)\left(1 - \sigma^{i^*}\right)\sigma^{z-i^*}}{\left(1 - \sigma^{z+1}\right)}$$

Substitute this into Eq. (11.16):

$$R_1 = (1+\sigma)\frac{(1+\sigma)(1-\sigma^{i^*})\sigma^{z+1-i^*}}{(1-\sigma^{z+1})} = (1+\sigma)\left(\frac{1-\sigma^{z+1}-\sigma^{z+1-i^*}+\sigma^{z+1}}{1-\sigma^{z+1}}\right)$$

$$= (1+\sigma)\left(\frac{1-\sigma^{z+1-i^*}}{1-\sigma^{z+1}}\right)$$

and now into equation $F_f(x) = KR_1$:

$$F_f(x) = KR_1 = \frac{1}{1+\sigma}R_1 = \frac{1-\sigma^{z+1-i^*}}{1-\sigma^{z+1}}$$

for example:

$$F_f(x) = \frac{1-\sigma^{z+1-i^*}}{1-\sigma^{z+1}}. \tag{11.17}$$

This is a relation for calculating the extent of fractional extraction of j size class into the fine product. The function $F_f(x)$ is not determined for $K = 0.5$ and $K = 0$. We examine these two cases:

1. $K = 0.5$, then $\sigma = 1$. Following l'Hôtital's rule: $\lim_{\sigma\to 1}\frac{1-\sigma^{z+1-i^*}}{1-\sigma^{z+1}} = \lim_{\sigma\to 1}\frac{(1-\sigma^{z+1-i^*})'}{(1-\sigma^{\#+1})'} = \frac{z+1-i^*}{z+1}$.
2. $K = 0$, then $\sigma = (1-0)/0$, hence, $\lim_{\sigma\to 1}\frac{1-\sigma^{z+1-i^*}}{1-\sigma^{z+1}} = 0$, since $z+1-i^* < z+1$ always holds. Then we finally obtain:

$$F_f(x) = \left|\begin{array}{ll} \dfrac{1-\sigma^{z+1-i^*}}{1-\sigma^{z+1}} & K \neq 0.5 \\ \dfrac{z+1-i^*}{z+1} & K = 0.5 \\ 0 & K = 0 \end{array}\right. . \tag{11.18}$$

To find the quantity of material of j size class coming out into the fine product, one should apply the formula $F_f(x) = r_f/r_s$ (i.e. $r_f = F_f r_s$), where r_s is known from the material composition. We can also estimate the narrow class output into the coarse product $r_{c,j} = r_{s,j} - r_{f,j}$.

3. Analysis of a mathematical model of a regular cascade

Principal properties of the obtained function $F_f(\sigma)$ can be formulated as follows:

1. According to the dependence Eq. (11.18), $F_f(\sigma)$ always exceeds zero with σ varying from 0 to ∞.
2. The range of values from 0 to 1 limits the change in $F_f(\sigma)$. $F_f(\sigma) = 1$ for $\sigma = 0$, and $F_f(\sigma) = 0$ for $\sigma \to \infty$.
3. The function $F_f(\sigma)$ is continuous and differentiable over the entire range of the distribution coefficient variation.
4. According to Eq. (11.18), $F_f(\sigma)$ is a single-valued function.

5. It can be shown that Eq. (11.18) can be reduced to Eq. (11.19):

$$F_f(\sigma)^{z+1} - \sigma^{z+1-i^*} + \left[1 - F_f(\sigma)\right] = 0 \tag{11.19}$$

with real coefficients. By the Descartes theorem, Eq. (11.19) should have two real roots or no roots at all. But because unity is an identity root of Eq. (11.19), the latter has a second real root in a particular case, which can also equal unity. Because this formula is used for $\sigma \neq 1$, only one real value $\sigma \neq 1$ corresponds to any fixed value $F_f(\sigma)$ of Eq. (11.19).

6. It follows from items 4 and 5, that $F_f(\sigma)$ is a monotonic function, which well agrees with physics of the process. Taking item 2 into account, $F_f(\sigma)$ is a monotonically decreasing function. Hence, $\partial F_f(\sigma)/\partial\sigma \leq 0$ over the entire range of the distribution factor variation. Passing to the distribution factor K, we obtain, respectively:

$$\frac{\partial F_f(\sigma)}{\partial K} = \frac{\partial F_f(\sigma)}{\partial\sigma}\frac{\partial\sigma}{\partial K} = -\frac{1}{K^2}\frac{F_f(\sigma)}{\partial\sigma} \geq 0$$

over the entire range of this factor variation. $\partial F_f(\sigma)/\partial\sigma$ or $\partial F_f(\sigma)/\partial K$ equal zero in limiting cases only, for $\sigma = 0$ or ∞ and $K = 1$ or 0, respectively. Both these values are marginal for σ or K and hardly of interest for the classification process.

The location of fractional separation curve in the coordinate system provides a complete characteristic of classification process quality – the steeper $F_f(\sigma)$ curve, the higher separation quality.

We determine the limit to which fractional extraction tends at a sufficiently large number of apparatus stages. For this purpose, we reduce Eq. (11.18) to the following expressions:

$$F_f(x)^{(i^*)}_{(z)/K\neq 0.5} = \frac{1 - \left(\frac{1-K}{K}\right)^{z+1-i^*}}{1 - \left(\frac{1-K}{K}\right)^{z+1}}$$

or:

$$F_f(x)^{(i^*)}_{(z)/K\neq 0.5} = \frac{\left(\frac{K}{1-K}\right)^{z+1} - \left(\frac{K}{1-K}\right)^{i^*}}{\left(\frac{K}{1-K}\right)^{z=1} - 1},$$

$$F_f(x)^{(i^*)}_{(z)/K\neq 0.5} = \frac{z+1-i^*}{z+1} = 1 - \frac{i}{z+1}.$$

We examine possible alternatives:

$$1)\quad K > 0.5;\quad \frac{1-K}{K} < 1$$

In this case:

$$F_f(x)^{(i^*)}_{(z)/z\to\infty} = \lim_{z\to\infty}\frac{1 - \left(\frac{1-K}{K}\right)^{z+1-i^*}}{1 - \left(\frac{1-K}{K}\right)^{z+1}} = \lim_{z\to\infty}\left[1 - \left(\frac{1-K}{K}\right)^{z+1-i^*}\right] = 1 - \lim_{z\to\infty}\left(\frac{1-K}{K}\right)^{z+1-i^*}$$

$$2)\quad K = 0.5.$$

In this case:

$$F_f(x)^{(i^*)}_{(z)z\to\infty} = 1 - \lim_{z\to\infty} \frac{i^*(z)}{z+1}.$$

$$3)\quad K < 0.5;\quad \frac{K}{1-K} < 1$$

$$F_f(x)^{(i^*)}_{(z)z\to\infty} = \lim_{z\to\infty} \frac{\left(\frac{K}{1-K}\right)^{z+1} - \left(\frac{K}{1-K}\right)^{i^*}}{\left(\frac{K}{1-K}\right)^{z+1} - 1} = \lim_{z\to\infty} \left(\frac{K}{1-K}\right)^{i^*}$$

For instance, in case of material feed into the apparatus from above ($i^* = 1$), we obtain for $K \geq 0.5$:

$$F_f(K)_{z\to\infty} = 1 - \lim_{z\to\infty} \left(\frac{1-K}{K}\right)^z = 1;$$

and for $K \leq 0.5$:

$$F_f(K)_{z\to\infty} = \lim_{z\to\infty} \left(\frac{K}{1-K}\right) \frac{K}{1-K}.$$

In case of symmetrical feed of the material in the middle of the apparatus ($i^* = (z+1)/2$) for $K > 0.5$:

$$F_f(K)_{z\to\infty} = 1 - \lim_{z\to\infty} \left(\frac{1-K}{K}\right)^{(z+1)/2} = 1;$$

for $K = 0.5$:

$$F_f(K)_{z\to\infty} = \lim_{z\to\infty} \frac{z+1}{2(z+1)} = 0.5 \quad \text{invariant with respect to } z;$$

for $K \leq 0.5$:

$$F_f(K)_{z\to\infty} = \lim_{z\to\infty} \left(\frac{1-K}{K}\right)^{(z+1)/2} = 0$$

In case of material feed into the apparatus from below ($i^* = z$), for $K \geq 0.5$:

$$F_f(K)_{z\to\infty} = 1 - \lim_{z\to\infty} \left(\frac{1-K}{K}\right) = 1 - \frac{1-K}{K} = \frac{2K-1}{K};$$

and for $K \leq 0.5$:

$$F_f(K)_{z\to\infty} = \lim_{z\to\infty} \left(\frac{K}{1-K}\right)^z = 0$$

Thus, boundary-size grain can be specified in advance only in case of feed in the middle, irrespective of the number of sections of the apparatus. It should have $K = 0.5$, because in compliance with the process optimum, boundary-size grain extraction upward or downward is 50% (for apparatuses with an arbitrary number of stages).

To find boundary-size grain defined by distribution factor $K \neq 1(w,d)$ for any other place of material feed, the following equation should be solved with respect to K_m:

$$\frac{1-\left(\frac{1-K_m}{k_m}\right)^{z+1-i^*}}{1-\left(\frac{1-K_m}{K_m}\right)^{z=1}} = 0.5. \tag{11.20}$$

It clearly shows the dependence of distribution factor on the number of stages and place of material feed into the apparatus $K_m = f(z, i^*)$:

$$d_{m.opt} = f(w, z, i^*).$$

We derive a general expression for the first derivative of fractional extraction with respect to the distribution factor:

$$[F_f(K)] = \left(\frac{1-\sigma^n}{1-\sigma^m}\right)\frac{d\sigma}{dK},$$

where:

$$\frac{1-K}{K} = \sigma; \quad z+1-i^* = n; \quad z+1 = m;$$

$$\frac{d\sigma}{dK} = \left(\frac{1-K}{K}\right) = \frac{-K-(1-K)}{K^2} = -\frac{1}{K^2};$$

$$[F_f(K)] = \frac{-n\sigma^{n-1}(1-\sigma^m) + m\sigma^{m-1}(1-\sigma^n)}{(1-\sigma^m)^2}\left(-\frac{1}{K^2}\right),$$

where $\sigma = 1$; $K = 0.5$; $F_m = 0/0$.

To remove the uncertainty, we apply l'Hôtital's rule:

$$[F_f(K)_{\sigma=1}] = \lim_{\sigma\to 1; K\to 0,5} \frac{n-\sigma^{n-1}(1-\sigma^m) - m\sigma^{m-1}(1-\sigma^n)}{(1-\sigma^m)^2}\frac{1}{K^2}$$

$$= 4\lim_{\sigma\to 1}\frac{[n\sigma^{n-m}(1-\sigma^m) - m(1-\sigma^n)]}{[(1-\sigma^m)^2]} = 4\lim_{\sigma\to 1}\frac{n(n-m)\sigma^{n-m-1}}{-2m\sigma^{m-1}}$$

$$= -\frac{2n(n=m)}{m} = \frac{2(z+1-i^*)(-i^*)}{z+1} = \frac{2i^*(z+1-i^*)}{z+1}.$$

This is the maximum value of $tg\alpha$– slope of the tangent of fractional extraction curve for i^*-feed. The highest value of the curve slope angle for various places of material feed into the apparatus can be found from the condition:

$$\frac{d}{di^*}[F_f(x)]_{K=0,5} = 0;$$

$$\frac{d}{di^*}\left[\frac{\partial F_f(x)}{\partial K}\right]_{K=0,5} = \frac{2(z+1-i^*) - 2i^*}{z+1} = 0.$$

Consequently:

$$2(z+1) - 4i^* = 0,$$

whence:

$$i^* = \frac{z+1}{2}.$$

Thus, a cascade apparatus with symmetrical material feed in the middle possesses the highest separation capacity.

We can emphasise once more that the obtained results can be successfully applied to cascade separation processes of various nature, such as adsorption, rectification, extraction, isotopes separation, etc. However, processes of different nature will have different mechanisms of the formation of mono-components distribution factor value.

4. Separation at cyclic feed of initial material into cascade apparatus

Until now, we have considered the process of bulk material separation in a cascade apparatus and material feed during the same time intervals $\Delta\tau$. Now we examine the case of material feed into the apparatus after each $p\Delta\tau$ steps, p being a non-negative integer.

For the case of material separation and feed into the apparatus realised during equal time intervals, we have derived the following iterative matrix relation:

$$r_j^m = A r_j^{m-1} + b$$

$$A = \left| \begin{array}{cccccc} 0 & K_J & 0 & \dots & 0 & 0 \\ 1-K_j & 0 & K_j & \dots & 0 & 0 \\ 0 & 1-K_j & 0 & K_j & \dots & 0 \\ 0 & 0 & \cdot & \cdot & \cdot & \cdot \\ \cdot & \cdot & \cdot & \cdot & \cdot & K_j \\ 0 & 0 & 0 & 0 & 1-K_j & 0 \end{array} \right| \quad b = \left| \begin{array}{c} 0 \\ 0 \\ \cdot \\ \gamma \\ \cdot \\ 0 \end{array} \right| \begin{array}{c} \\ \\ i \\ \leftarrow \\ \\ \\ \end{array}$$

where γ is the material quantity in one portion fed to the apparatus, $r_j^m = (r_{1j}^m, r_{2j}^m, r_{3j}^m, \ldots, r_{zj}^m)$, where r_{ij}^m is the quantity of material of size class j at the apparatus stage number i during the m-th separation act.

When the material is fed into the apparatus after each p separation acts, the iterative matrix relation has the following form:

$$r_j^m = A r_j^{m-1} + b$$

$$r_j^{m+1} = A r_j^m$$

$$\dots$$

$$r_j^{m+p-1} = A r_j^{m+p-2}$$

$$r_j^{m+p} = A r_j^{m+p-1} + b$$

etc.

We define a new iterative process reflecting the state of material distribution after each p steps:

$$r_j^0 = u_j^0, r_j^\rho = u_j^1, r_j^{2\rho} = u_j^2, \ldots, r_j^{n\rho} = u_j^n, \cdots\cdots$$

Taking this into account:

$$\begin{aligned} u_j^1 &= A^\rho u_j^0 + b \\ u_j^2 &= A^\rho u_j^1 + b \\ &\cdots \\ u_j^n &= A^\rho u_j^{n-1} + b \end{aligned}$$

It has been proved that all eigenvalues of the matrix A are less than 1 modulo, and, hence, A^ρ possesses the same property. Then, the process is reduced to $u_j^n \underset{n\to\infty}{\to} u_j$, where u_j is the solution of iterative matrix equation $u_j = A^\rho u_j + b$.

Hence, we can conclude that the initial process r_j^n is cyclic, and each subsequence $r_j^{n_k}$ converges (where n_k is divisible by p).

We call such process in its limiting state stationary-cyclic. We denote the quantity of material of size class j at the separation stage i during p separation acts in stationary-cyclic state as follows (omitting the index j for convenience): $r_i^{(1)}, r_i^{(2)}, \ldots, r_i^{(\rho)}$. Note that $r_i^{(m)} = r_i^{(m+\rho)}$ for any non-negative integer m.

We determine $r_i = r_i^{(1)} + r_i^{(2)} + \ldots r_i^{(\rho)}$ (i.e. r_i is the total quantity of material of size class j passing through stage i during p separation acts). We have proved that r_i is constant.

Let R_i be a quantity of material of size class j passing through stage i during p separation acts normalised to its quantity in one portion of material fed into the apparatus.

Clearly, the value of fractional extraction of material of size class j during p separation acts into fine product is:

$$F_f(x) = KR_1$$

and into coarse product:

$$1 - F_f(x) = (1 - K)R_z$$

Quantities of material entering and coming out of stage i during p separation acts are equal and, hence, balance equations derived for a stationary process holds, for example:

$$\begin{gathered} (1-K)R_i + KR_i = (1-K)R_{i-1} + KR_{i+1} \quad i \neq 1; z; i^\bullet - 1; i^\bullet + 1 \\ (1-K)R_{i^\bullet-1} + KR_{i^\bullet-1} = (1-K)R_{i^\bullet-2} + KR_{i^\bullet} + K \\ (1-K)R_{i^\bullet+1} + KR_{i^\bullet+1} = (1-K)R_{i^\bullet} + KR_{i^\bullet+2} + (1-K) \\ R_1 = KR_2 \\ R_z = (1-K)R_{z-1}. \end{gathered}$$

Hence, the formula for estimating fractional extraction degree during p separation acts is the same as for a stationary process, namely:

$$F_f = \begin{vmatrix} \dfrac{1-\sigma^{z+1-i^{\bullet}}}{1-\sigma^{z+1}} & k\neq 0.5 \\ \dfrac{z+1-i^{\bullet}}{z+1} & k=0.5 \\ 0 & k=0 \end{vmatrix}. \tag{11.21}$$

Fractional extraction of material of size class j during p separation acts can also be estimated as a ratio of the quantity of this size class coming out into the fine product during p separation acts to its quantity in one portion of material fed into the apparatus.

The obtained relation for fractional extraction remains constant during the entire process, which follows from balance equations.

The only difference between a stationary-cyclic process and a stationary process is that the former has the number of separation acts multiple to p if it is finite, and all computations should be performed, respectively.

5. Absorbing Markov chain in cascade classification of bulk materials

The process under study is similar to a random travel upward and downward with upward transition with the probability K and downward transition with the probability $1-K$ and two absorbing boundary states. Therefore, redistribution of particles of a fixed class in an apparatus of z stages can be represented by an absorbing Markov chain having a matrix of the following form:

$$\begin{vmatrix} 1 & 0 & 0 & 0 & 0 & 0 & \dots & 0 \\ K & 0 & 1-k & 0 & 0 & 0 & \dots & 0 \\ 0 & K & 0 & 1-k & 0 & 0 & \dots & 0 \\ \cdot & \cdot & \cdot & \cdot & \cdot & \cdot & \cdot & \cdot \\ \cdot & \cdot & \cdot & \cdot & \cdot & \cdot & \cdot & \cdot \\ 0 & 0 & \dots & 0 & K & 0 & 1-k & 0 \\ 0 & 0 & \dots & 0 & 0 & K & 0 & 1-k \\ 0 & 0 & 0 & 0 & 0 & \dots & 0 & 1 \end{vmatrix}$$

This matrix has $(z+2)$ rows and columns, the first and the last columns are outputs into coarse and fine products, and the remaining columns are probabilities of particles transitions between the apparatus stages.

We transform this matrix into a more convenient canonic form combining all ergodic (absorbing) states into one group and all nonrecurrent states into another. We obtain z nonrecurrent states (corresponding to the number of apparatus stages) and two ergodic states (corresponding to coarse and fine products). Then the canonic form is as follows:

$$
\begin{array}{c}
\qquad\qquad \downarrow I \qquad\qquad\qquad\qquad\qquad \downarrow O \\
R \rightarrow \left|\begin{array}{cc|ccccccc}
1 & 0 & 0 & 0 & \cdot & \cdot & \cdots & 0 & 0 \\
0 & 1 & 0 & 0 & \cdot & \cdot & \cdots & 0 & 0 \\
\hline
K & 0 & 0 & 1-K & 0 & \cdot & \cdot & \cdots & 0 \\
0 & 0 & K & 0 & 1-k & 0 & \cdot & \cdots & 0 \\
0 & 0 & 0 & K & 0 & 1-K & 0 & \cdots & 0 \\
\cdot & \cdot & \cdot & \cdot & \cdot & \cdot & \cdot & \cdot & \cdot \\
0 & 0 & 0 & \cdots & 0 & k & 0 & 1-K & 0 \\
0 & 0 & 0 & 0 & \cdots & 0 & K & 0 & 1-K \\
0 & 1-K & 0 & 0 & 0 & \cdots & 0 & k & 0
\end{array}\right| \leftarrow Q
\end{array}
$$

Here, the region O is entirely composed of zeroes, the submatrix Q (with the dimension $z \times z$) describes the particle behaviour before its exit of the apparatus (from the multitude of nonrecurrent states), the submatrix R with the dimension $(z \times 2)$ corresponds to transitions from the apparatus into coarse and fine products (from nonrecurrent to ergodic state), and the submatrix I with the dimension (2×2) refers to the process after the particle exit from the apparatus (after the particle reaches ergodic state). The probability of reaching one of such states in an absorbing chain tends to 1. Hence, we can state that sooner or later, a particle will reach a certain absorbing state (i.e. come out into one of separation products), with the probability of 1.

For any absorbing chain, Q^n tends to zero, and $I - Q$ is reversible, with $(I - Q)^{-1} = \sum_{K=0}^{\infty} Q^K$. For an absorbing Markov chain, the matrix $N = (I - Q)^{-1}$ is called a fundamental matrix.

We denote the function equal to the total number of redistribution acts at the stage j by n_j. Then we can state that the mathematical expectation for a particle located at stage i at the initial moment of time to pass to stage j after n_j redistribution acts is the coordinate ij of the matrix N, for example:

$$\{E_i(n_j)\} = N. \tag{11.22}$$

Hence, average time of a particle stay at a given stage is always finite, and these average times are just the matrix N.

Each particle gets into the apparatus through stage $i^{\bullet}$, and using Eq. (11.22) we can determine the average time of particle stay at each stage.

We introduce the following notations:

$$N_2 = N(2N_{dg} - I) - N_{sq}$$

N_2 is a matrix of the size $z \times z$, where N_{dg} corresponds to a matrix whose diagonal equals the diagonal of N matrix, and all other elements are zero, while N_{sq} corresponds to a matrix whose elements are squares of elements of N matrix. Clearly, in a general case, $A^2 \neq A_{sq}$ is valid for any matrix A. An equality corresponds to diagonal matrices only.

$$B = NR \quad \text{is a matrix of the size} \quad z \times 2.$$

$$\tau = N\xi$$

where ξ is a column vector with all elements equal to 1.

$$\tau_2 = (2N - I)\tau - \tau_{sq}$$

One can assert that, for absorbing Markov chains, the probability for a particle located in stage i at the initial moment of time to pass to stage j after n_j redistribution acts corresponds to the coordinate ij of the matrix N_2, for example:

$$\{D_i(n_j)\} = N_2. \tag{11.23}$$

Let the function T equal the total time (number of redistribution acts including the initial position) of the particle stay inside the apparatus before its exit. The quantity T shows the number of Markov process steps to be made before the particle leaves the apparatus.

If at the initial moment of time a particle is at the stage i, mathematical expectation of the number of redistribution acts before the particle gets into coarse or fine product is the i-th coordinate of the vector τ, and its dispersion is the i-th coordinate of the vector τ_2, for example:

$$\{E_i(T)\} = \tau$$

$$\{D_i(T)\} = \tau_2$$

In the process of separation, a particle is fed into the apparatus through stage i^*, and, hence, the average time of particle stay in the apparatus (the number of redistribution acts) and deviation from this time can be determined.

Let b_{ij} be the probability of the fact that a particle staying at the stage i gets into coarse or fine product ($j = 1$ for fine product and $j = 2$ for coarse one). Then:

$$\{\mathrm{b_{ij}}\} = \mathrm{B} = \mathrm{NR}$$

Hence, it is possible to control the probability of particle exit into the required product by changing the place of its feed into the apparatus. This allows one to estimate any version of cascade separator design.

6. Mathematical model of cascade separation with arbitrary distribution coefficients

There exist apparatuses operating in a nonuniform regime over the height of separating column, such as a classifier with a variable cross-section along the height (e.g. conical), with additional feed or discharge of continuous fluid flows, etc. Such apparatuses are characterised by a nonuniform velocity field along the height and, hence, by different distribution coefficients of monofractions in different cross-sections,

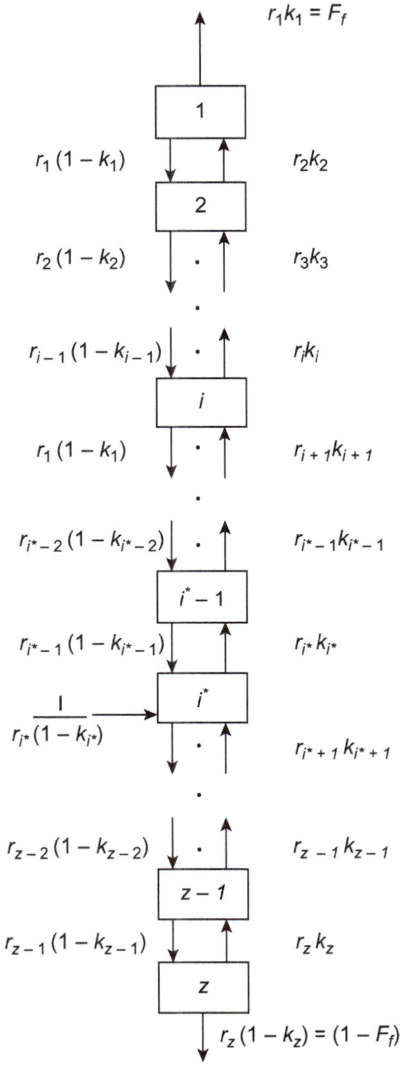

FIGURE 11.4 Scheme of a nonuniform cascade.

and can be considered as a totally nonuniform cascade of separating elements. Figure 11.4 shows a scheme of a totally nonuniform cascade in case of equilibrium principle of the elements operation.

Let $K_1 \ldots K_z$ be distribution coefficients at each stage, respectively. The examination of mono-fraction flows in a nonuniform cascade (Figure 11.4) shows that the following is valid for material balance of the entire column:

$$r_1K_1 + r_z(1 - K_z) = 1 \tag{11.24}$$

The condition of material balance for the upper part of the column from the first to the i-th element is:

$$r_1K_1 + r_i(1 - K_i) = r_{i+1}K_{i+1} \tag{11.25}$$

$$i \leq i^* - 1.$$

Hence:

$$r_{i+1}K_{i+1} - r_i(1 - K_i) = F_f \tag{11.26}$$
$$i \le i^\bullet - 1.$$

Similarly, for the part of the column of the lower branch of the apparatus:

$$r_z(1 - K_z) + r_i K_i = r_{i-1}(1 - K_{i-1}) \tag{11.27}$$
$$i \le i^\bullet + 1$$

whence we obtain:

$$r_{i-1}(1 - K_{i-1}) = 1 - F_f$$
$$i \le i^\bullet + 1.$$

Changing indices in recurrent Eq. (11.26), we obtain:

$$\begin{cases} r_iK_i - r_{i-1}(1 - K_{i-1}) = F_f; \quad (i \le i^\bullet); \\ r_i(1 - K_i) - r_{i+1}K_{i+1} = 1 - F_f; \quad (i \le i^\bullet) \end{cases} \tag{11.28}$$

For $i{=}i^\bullet$, Eq. (11.28) expresses the material balance conditions. Using Eq. (11.28) and boundary conditions for the upper branch of the column, we derive a system of equations:

$$\begin{cases} r_1K_1 = F_f \\ r_2K_2 = r_1(1 - K_1) + F_f \\ r_3K_3 = r_2(1 - K_2) + F_f \\ \cdots \\ r_iK_i = r_{i-1}(1 - K_{i-1}) + F_f \\ \cdots \\ r_{i^\bullet}K_{i^\bullet} = r_{i^\bullet-1}(1 - K_{i^\bullet-1}) + F_f \end{cases} \tag{11.29}$$

Using the method of successive substitutions, we transform the system in Eq. (11.29) into the following form ($i \le i^\bullet$):

$$\begin{cases} r_1K_1 = F_F \\ r_2K_2 = F_f\sigma_1 + F_f \\ r_3K_3 = \sigma_2(F_f\sigma_1 + F_f) + F_f \\ \quad = \sigma_2\sigma_1F_f + \sigma_2F_f + F_f \\ r_4K_4 = \sigma_3[\sigma_2(F_f\sigma_1 + F_f) + F_f] + F_f \\ \quad = \sigma_3\sigma_2\sigma_1F_f + \sigma_3\sigma_2F_f + \sigma_3F_f \\ r_5K_5 = \sigma_4\sigma_3\sigma_2\sigma_1F_f + \sigma_4\sigma_3\sigma_2F_f \\ \quad +\sigma_4\sigma_3F_f + \sigma_4F_f + F_f \\ \cdots \end{cases}$$

Hence,

$$r_iK_i = F_f\left(\sum_{l=1}^{i-1}\prod_{l}^{i-1}\sigma_l + 1\right)$$

or

$$r_i K_i = \frac{F_f}{\sigma_i} \sum_{l=-1}^{i} \prod_{l}^{i} \sigma_l; \quad (\leq i \leq i^{\bullet}) \tag{11.30}$$

is valid for the i-th element, where $\sigma_i = (1 - K_i)/K_i$ is the distribution index for the i-th element of the cascade.

Taking into account Eq. (11.30) for i^* element, we obtain:

$$r_{i^{\bullet}} K_{i^{\bullet}} = \frac{F_f}{\sigma_{i^{\bullet}}} \sum_{l=1}^{i^{\bullet}} \prod_{l}^{i^{\bullet}} \sigma_l \tag{11.31}$$

Using the boundary condition and Eq. (11.28), monofraction flows directed upwards are determined for the lower cascade branch ($i \geq i^{\bullet}$)

$$\begin{cases} r_{i^{\bullet}+1} K_{i^{\bullet}+1} = r_{i^{\bullet}}(1 - K_{i^{\bullet}}) - (1 - F_f) \\ r_{i^{\bullet}+2} K_{i^{\bullet}+2} = r_{i^{\bullet}+1}(1 - K_{i^{\bullet}+1}) - (1 - F_f) \\ r_{i^{\bullet}+3} K_{i^{\bullet}+3} = r_{i^{\bullet}+2}(1 - K_{i^{\bullet}+2}) - (1 - F_f) \\ \cdots \\ \cdots \\ \cdots \\ r_z K_z = r_{z-1}(1 - K_{z-1}) - (1 - F_f) \\ r_z K_z \sigma_z = 1 - F_f \end{cases} \tag{11.32}$$

Taking into account Eq. (11.31), the system of Eq. (11.32) is transformed into:

$$r_{i^{\bullet}+1} K_{i^{\bullet}+1} = F_f \sum_{l=1}^{i^{\bullet}} \prod_{l}^{i^{\bullet}} \sigma_l - (1 - F_f);$$

$$r_{i^{\bullet}+2} K_{i^{\bullet}+2} = \sigma_{i^{\bullet}+1} F_f \sum_{l=1}^{i^{\bullet}} \prod_{l}^{i^{\bullet}} \sigma_l - \sigma_{i^{\bullet}+1}(1 - F_f) - (1 - F_f);$$

$$r_{i^{\bullet}+3} K_{i^{\bullet}+3} = \sigma_{i^{\bullet}+2}\sigma_{i^{\bullet}+1} F_f \sum_{l=1}^{i^{\bullet}} \prod_{l}^{i^{\bullet}} \sigma_l - \sigma_{i^{\bullet}+2}\sigma_{i^{\bullet}+1}(1 - F_f) - \sigma_{i^{\bullet}+2}(1 - F_f) - (1 - F_f);$$

$$\cdots$$

$$\cdots$$

The obtained expressions allow writing the following for an arbitrary stage ($i \geq i^{\bullet}$):

$$r_i K_i = i - \prod_{i^{\bullet}+1}^{i-1} \sigma_l \cdot F_f \sum_{l=1}^{i^{\bullet}} \prod_{l}^{i^{\bullet}} \sigma_l - (1 - F_f)\left(1 + \sum_{l=i^{\bullet}+1}^{i-1} \prod_{l}^{i-1} \sigma_l\right)$$

We transform the latter expression:

$$r_iK_i = \frac{F_f}{\sigma_i}\sum_{l=1}^{i^*}\prod_{l}^{i}\sigma_l - \frac{(1-F_f)}{\sigma_i}\sum_{l=i^*+1}^{i}\prod_{l}^{i}\sigma_l$$

or

$$r_iK_i = \frac{F_f}{\sigma_i}\sum_{l=1}^{i}\prod_{l}^{i}\sigma_l - \frac{1}{\sigma_i}\sum_{l=i^*+1}^{i}\prod_{l}^{i}\sigma_l; \quad (i^* \le i \le z) \tag{11.33}$$

whence we obtain the following for the last element of the cascade:

$$r_zK_z = \frac{F_f}{\sigma_z}\sum_{l=1}^{z}\prod_{l}^{z}\sigma_l - \frac{1}{\sigma_z}\sum_{l=i^*+1}^{z}\prod_{l\sigma_l}^{z}$$

Taking Eq. (11.32) into account, we can write:

$$1 - F_f = F_f\sum_{l=1}^{z}\prod_{l}^{z}\sigma_l - \sum_{l=i^*+1}^{z}\prod_{l}^{z}\sigma_l$$

From the obtained relation, fractional extraction is expressed as:

$$F_f = \frac{1+\sum_{l=i^*+1}^{z}\prod_{l}^{z}\sigma_l}{1+\sum_{l=1}^{z}\prod_{l}^{z}\sigma_l}. \tag{11.34}$$

The obtained expression can be transformed by multiplying both the numerator and denominator by $\prod_{l=1}^{z}\frac{1}{\sigma_l}$, then we obtain:

$$F_f = \frac{\sum_{l=i^*}^{z}\prod_{1}^{l}\left(\frac{1}{\sigma}\right)_l}{1+\sum_{l=1}^{z}\prod_{1}^{l}\left(\frac{1}{\sigma}\right)_l} \tag{11.35}$$

According to Eq. (11.30), the total monofraction flow in an arbitrary element of the upper branch of the cascade is:

$$r_i = \frac{F_f}{(1-K_i)}\cdot\sum_{l=1}^{i}\prod_{l}^{i}\sigma_l; \quad (1 \le i \le i^*). \tag{11.36}$$

Similarly, from the expression in Eq. (11.33) for the lower branch:

$$r_i = \frac{1}{(1-K_i)}\left(F_f \sum_{l=1}^{i}\prod_{l}^{i}\sigma_l - \sum_{l=i^*+1}^{i}\prod_{l}^{i}\sigma^*\right); \quad (i^* \le i \le z). \tag{11.37}$$

Hence, we can write in the general form:

$$F_f = \frac{1+\sum_{n=i^*+1}^{z}\prod_{q=n}^{z}\sigma_q}{1+\sum_{n=1}^{z}\prod_{q=n}^{z}\sigma_q} = \frac{\sum_{n=i^*}^{z}\prod_{q=1}^{z}\sigma_q}{1+\sum_{n=1}^{z}\prod_{q=1}^{z}\sigma_q};$$

$$r_i = F_f\frac{(\sigma_i+1)}{\sigma_i}\sum_{n=1}^{i}\prod_{q=n}^{i}\sigma_q; \quad (1 \le i \le i^*);$$

$$r_i = \frac{(\sigma_i+1)}{\sigma_1}\left[F_f\sum_{n=1}^{i}\prod_{q=n}^{i}\sigma_q - \sum_{n=i^*}^{i}\prod_{q=n}^{i}\sigma_q\right]; \quad (i^* \le i \le z)$$

Make sure that Eqs. (11.35), (11.36), and (11.37) are valid for a regular cascade ($\sigma_i = \sigma$). In compliance with Eq. (11.35):

$$F_f = \frac{1+\sigma^{z-i^*}+\sigma^{z-i^*-1}+\ldots+\sigma^2+\sigma}{1+\sigma^z+\sigma^{z-1}+\ldots+\sigma^2+\sigma}$$

Using the formula of the sum of terms of geometric series with the denominator $\sigma \neq 1$, we obtain:

$$F_f = \frac{1-\sigma^{z+1-i^*}}{1-\sigma^{z+1}},$$

which corresponds to previously obtained expression.

7. Discrete model of critical regimes of vertical two-phase flows

A characteristic property of such flows is a nonstationary multidirectional motion of particles and their ability to change the motion direction many times during their stay in the channel. The direction can be altered at any point along the channel height. All this considerably complicates the analysis of the flow mechanism and of the solid phase distribution between two outputs of the process – upper and lower. Such processes are widely used in industrial practice, for instance, in fractionation of bulk materials according to particles sizes or densities.

We assume that a poly-fractional bulk material is fed into a vertical channel. Under the action of air flow, a part of this material goes out upward, and another part settles downward. The finest particles are completely carried out upward, and the coarsest particles settle downward. Particles of intermediate classes are shared by two outputs – those of fine and coarse products.

If the initial granulometric composition of the material is known, determining the composition of one of the outputs allows automatic determination of the product composition in another output. We assume that the fine product is the determining one.

To develop a model of the process of a critical two-phase flow, we define its boundary conditions.

1. The process model is discrete both in space and time.
   Spatial discreteness is determined by the following reasons. If we take a differential of the amount of material at a certain level of the channel height and then integrate over the entire height, it leads to nothing, because changes in the direction of the particles motion can start at any level.

   Therefore, we mentally divide the vertical channel along its height into a finite number of identical elements and call each element a stage. The idea of channel subdivision into stages consists in the assumption that changes in the direction of the particles occur only within a stage, and the latter exchanges particles with adjacent stages only. Material can be fed into such a flow from above, from below or into any stage.

   Assume that the channel under study consists of z stages (z being a natural number). We count the stages top-down and denote any stage by $i(1 \leq i \leq z)$. If we consider a certain stage i, the material located therein partially goes out into the overlying stage $(i-1)$ under the action of an ascending flow, while another part gets into the stage $(i+1)$ located below. At the same time, a part of the same material arrives from these stages at the stage i. We denote the stage of the initial material feed by i^*. Clearly, in a general case, $1 \leq i^* \leq z$. Time discreteness is stipulated by the fact that, putting aside a continuous temporal transformation of the vector of polyfractional mixture composition, the latter is considered only at certain fixed moments of time with the interval $\Delta\tau$. It means that this model examines instantaneous and simultaneous particles redistribution in all stages of the channel occurring at discrete moments of time with the same interval $\Delta\tau$. We call such distribution of polydisperse material a solitary separation act.
2. The model is related to the field of self-similar processes with respect to the average consumed material concentration μ in the flow. In the concentration range up to 2.5 kg/m^3, separation in the flow is independent of the solid phase content in it. This result is independent of concentration, the method of the material feed, mean flow velocity value or the number of stages.
3. Each narrow size class of particles is separated in such flows independently, irrespective of the presence of other classes of particles. Statements 2 and 3 are closely interconnected and experimentally established. Statement 3 allows us to pass from a polydisperse material to the examination of independent behaviour of each monofraction separately r_{ij}^n. Further, we omit the index n, and all our conclusions are related to one monofraction only.

   If we successively determine distributions of all narrow classes, we obtain complete information on the result of the process. Usually, this is an objective of any prediction. Each redistribution of a monofraction at discrete moments of time in all channel sections represents an isolated separation act (ISA).

   Here, we denote by ρ_{ij} the amount of monofraction in the i-th stage at the j-th ISA at a single initial feed of this fraction into the flow.

Respectively, r_{ij} is the total flow of this monofraction in the i-th stage during j ISAs. In the general case, at an arbitrary feed:

$$r_{ij} = \sum_{0}^{j} \rho_{ij}.$$

4. We examine an equilibrium model implying the conservation of material balance between the amounts of particles contained in and leaving any section during an arbitrary ISA.
5. We examine a stationary model only, which assumes that the number of particles entering a certain stage (i) equals the number of particles leaving this stage.
6. We assume that particles separation coefficient in any stage remains unchanged at an unchanged ascending flow velocity. Here, the separation coefficient k implies the portion of particles of a narrow size class passing into the overlying cascade section during an ISA (i.e. $k = (r_{i-1})/r_i$). For identical stages, the value $k = const$, which has been confirmed experimentally.

According to items 4 and 6, the distribution of an arbitrary monofraction in any section during the $(j + I)$-th ISA is expressed as follows (Figure 11.1(a)):

$$\rho_{ij} = \rho_{ij} \cdot k + \rho_{ij}(1 - k)$$

where ρ_{ij} is the number of particles of a fixed narrow size class staying in the i-th stage during the j-th ISA at a discrete moment of time; $\rho_{ij} \cdot k$ is the fraction of these particles that passed to the overlying $(i-1)$-th stage at the $(j+1)$-th ISA at a discrete moment of time; $\rho_{ij}(1-k)$ is the fraction of particles that passed to $(i+1)$-th stage.

Particles content in a fixed fraction in separate stages during an arbitrary number j of ISA can be expressed by a certain matrix of values.

We examine a single feed of the initial number of particles taken as unity (a portion). A general pattern of particles distribution in separate stages during an arbitrary number j of ISA can be expressed by the following matrix of values:

τ_j	τ_0	τ_1	τ_2	τ_3	τ_4	...	τ_j	...
j/i	0	1	2	3	4	...	j	...
1	ρ_{10}	ρ_{11}	ρ_{12}	ρ_{13}	ρ_{14}	...	$\rho_{i,j}$	
$\vdots$	$\vdots$	$\vdots$	$\vdots$	$\vdots$	$\vdots$	...	$\vdots$	...
$i^{\bullet} - 1$	0	k	0	$3k^2(1-k)$	$\rho_{i^{\bullet}-1,4}$	...	$\rho_{i^{\bullet}-1,j}$	...
$i^{\bullet} - \rightarrow 1$		0	$2k(1-k)$	0	$6k^2(1-k)^2$	...	$\rho_{i^{\bullet},j}$	...
$i^{\bullet} + 1$	0	$1-k$	0	$3k(1-k)^2$	$\rho_{i^{\bullet}+1,4}$	...	$\rho_{i^{\bullet}+1,j}$	...
$\vdots$	$\vdots$	$\vdots$	$\vdots$	$\vdots$	$\vdots$	...	$\vdots$	...
$z - 1$	$\rho_{z-1,0}$	$\rho_{z-1,1}$	$\rho_{z-1,2}$	$\rho_{z-1,3}$	$\rho_{z-1,4}$	...	$\rho_{z-1,j}$	...
z	$\rho_{z,0}$	$\rho_{z,1}$	$\rho_{z,2}$	$\rho_{z,3}$	$\rho_{z,4}$	...	$\rho_{z,j}$	...

where $\tau_j = j\Delta\tau$ are discrete moments of time.

In case of a single feed, we obtain an infinite matrix in the general form:

$$\rho_{ij}=\begin{bmatrix}\rho_{10} & \rho_{11} & \rho_{12} & \cdots & \rho_{ij} & \cdots\\ \rho_{20} & \rho_{21} & \rho_{22} & \cdots & \rho_{ij} & \cdots\\ \cdots & \cdots & \cdots & \cdots & \cdots & \cdots\\ \rho_{i0} & \rho_{i1} & \rho_{i2} & \cdots & \rho_{ij} & \cdots\\ \cdots & \cdots & \cdots & \cdots & \cdots & \cdots\\ \rho_{z0} & \rho_{z1} & \rho_{z2} & \cdots & \rho_{zj} & \cdots\end{bmatrix}.$$

It follows from the preceding distribution scheme that:

$$\rho_{ij}=\rho_{i-1,j-1}(1-k)+\rho_{i+1,j-1}k \tag{11.38}$$

under the initial conditions:

$$\left.\begin{aligned}\rho_{i0/i\neq i^{\bullet}}&=0\\ \rho_{i^{\bullet},0}&=1\end{aligned}\right\}$$

and boundary conditions:

$$\left.\begin{aligned}\rho_{1,j}&=\rho_{2,j-1}k\\ \rho_{z,j}&=\rho_{z-1,j-1}(1-k)\end{aligned}\right\}.$$

Theoretically, in the most general case, the number of elements in any line of the matrix $[\rho_{ij}]$ is infinite. However, in practice, after a certain period of time ($T_m=m\cdot\Delta\tau$), all the particles leave the apparatus passing into fine or coarse product, so that:

$$\rho_{ij/j\geq m+1}=0.$$

Thus, the initial infinite matrix can be restricted by m ISAs:

$$\left[\rho_{i_1\infty}\right]=\left[\rho_{i,m}\right]=\begin{bmatrix}\rho_{10} & \rho_{11} & \rho_{12} & \cdots & \rho_{1m}\\ \rho_{20} & \rho_{21} & \rho_{22} & \cdots & \rho_{2m}\\ \cdots & \cdots & \cdots & \cdots & \cdots\\ \rho_{z0} & \rho_{z1} & \rho_{z2} & \cdots & \rho_{zm}\end{bmatrix}. \tag{11.39}$$

In the general case, the distribution r_{ij} of a narrow class of particles along the channel height depends on the number of stages, distribution coefficient, initial and boundary conditions, fixed moment of time, site and method of the initial material feed.

A discrete model presupposes the initial product feed in equal portions at equal time intervals $\Delta T=t\Delta\tau(t=1,2,3,\ldots m)$. Evidently, $t\geq 1$, because otherwise the number of feeds would exceed the number of ISAs, which would lead to unlimited accumulation of particles. Thus, particles content in the general case is expressed as:

$$r_{ij}=f(k,z,i^{\bullet},i,j,t). \tag{11.40}$$

The realisation of various methods of the initial material feed leads to the following main regimes of feeding:

1. A continuous regime (within the bounds of a discrete model) is realised at the step of $t=1$. The interval between subsequent separate feeds is $\Delta\tau$. Since the moment of time $\tau_j=\tau_m$, the distribution becomes stationary.

   The discontinuity of the moments of time with the step $\tau=j\Delta\tau$ corresponds to:

$$j=0,1,2\ldots$$

   Then the amount of a monofraction at a given moment of time j (during j ISAs) is described by a relationship:

$$r_{ij}=\sum_{\gamma_0}^{j}\Big(\rho_{i(j-j_0)}. \tag{11.41}$$

The latter can be written as:

$$r_{ij}=\rho_{ij}+\rho_{i,j-1}+\cdots+\rho_{i,0}=\sum_{0}^{j}\rho_{ij}.$$

A similar expression is valid in the matrix form of Eq. (11.39) type:

$$\left[r_{ij}\right]=\sum_{0}^{j}\left[\rho_{ij}\right]. \tag{11.42}$$

Because $\rho_{ij/j\geq m+1}=0$, we can write:

$$r_{ij/j\geq m}=\sum_{0}^{m}\rho_{ij}.$$

Thus, since the moment of time $\tau_j=\tau_m$, a stationary distribution regime ensues, which is characterised by an invariable particles flow over the sections:

$$r_i=r_{ij/\gamma\geq m}=\sum_{0}^{m}\rho_{ij}=const. \tag{11.43}$$

Hence, in this case, the number of dependent variables decreases in comparison with Eq. (11.41):

$$r_i=f(r,z,i^*,i). \tag{11.44}$$

Using Eq. (11.38), we can describe an equilibrium particles flow in any section of the channel. For this purpose, we sum up the expression:

$$\rho_{i;j+1}=\rho_{i-1;j}(1-k)+\rho_{i+1;j}k \tag{11.45}$$

over j at its variation from 0 to m:

$$\sum_0^m \rho_{i;j+1} = (1-k)\sum_0^m \rho_{i-1;j} + k\sum_o^m \rho_{i+1;j}. \tag{11.46}$$

Since at $i \neq i^{\bullet}$:

$$\sum_0^m \rho_{i;j+1} = \sum_0^m \rho_{i;j},$$

because $\rho_{i0} = \rho_{i;m+1} = 0$, we obtain:

$$\sum_0^m \rho_{i;j} = (1-k)\sum_0^m \rho_{i-1;j} + k\sum_0^m \rho_{i+1;j}. \tag{11.47}$$

In accordance with Eq. (11.43), this equation can be written in the form:

$$r_i = r_{i-1}(1-k) + r_{i+1}k. \tag{11.48}$$

Because Eq. (11.48) is stipulated by the model equilibrium and stationary state, the same dependence can be derived from the equation of material balance of a stage:

$$r_i = r_i(1-k) + r_i(k) = r_{i-1}(1-k) + r_{i+1}k \tag{11.49}$$

at the boundary condition of the material feed into the flow:

$$r_{i^{\bullet}} = 1 + r_{i^{\bullet}-1}(1-k) + r_{i^{\bullet}+1}k, \tag{11.50}$$

and the boundary conditions of material outlet of the channel to both sides:

$$\left.\begin{array}{r} r_1 = r_2 k \\ r_z = r_{z-1}(1-k) \end{array}\right\}. \tag{11.51}$$

2. Periodic regime of feeding with the step $t = m + 1$.

It means that each new unit portion of a monofraction enters the flow when all the particles of the previous portion have left the channel.

The corresponding moments of the material feed are:

$$j = 0; (m+1), 2(m+1), 3(m+1)....$$

For this case, the following is valid:

$$r_{ij} = \rho_{ij}, \tag{11.52}$$

since individual feeds do not overlap.

Hence, it is obvious that within any cycle, the respective flows are repeated (the process is stationary within the bounds of a cycle):

$$\rho_{ij} = \rho_{ij+s(m+1)} = r_{i,j+s(m+1)} = r_{i,j+st};(s = 0, 1, 2, 3...). \tag{11.53}$$

3. Periodic regime of feeding with the step $1 < t < m+1$.
   In contrast to a continuous feed, here moments of feed are:

$$j = 0; t; 2t; 3t\ldots.$$

Besides, after a time interval $\tau = t \cdot \Delta\tau$, the total flow in any stage is repeated (i.e. starting from the moment of time $\tau_j = \tau_{m+1-t}$), a cyclically stationary regime ensues within the bounds of one feed step.

By analogy with Eq. (11.41), in this case, we can write:

$$r_{ij} = \sum_{s=0}^{c} \rho_{i;st-j_0} = \sum_{s=0}^{c} \rho_{i;st}. \tag{11.54}$$

Summing up, we can conclude that at any method of feeding a single monofraction into the flow, the process is stationary (the content in stages is repeated after every step).

Thus, starting from the moment of time $\tau_i = \tau_m$, a stationary regime ensues, which is characterised by an unchanged content of particles in stages:

$$r_{ij/_{j \geq m}} = r_i = \sum_{j=0}^{m} \rho_{ij} = const. \tag{11.55}$$

Hence, in this regime $r_i = f(k; z; i^{\bullet}; i)$.

Using Eq. (11.38), we can express the equilibrium content of particles in an arbitrary section through their contents in neighbouring sections. For this purpose, we summarize the expression:

$$\rho_{i,j+1} = \rho_{i-1,j}(1-k) + \rho_{i+1,j}k; \quad \text{over } j \text{ at } j \text{ varying from 0 to } m$$

$$\sum_{j=0}^{m} \rho_{i,j+1} = \sum_{j=0}^{m} \rho_{i-1,j}(1-k) + \sum_{j=0}^{m} \rho_{i+1,j}k.$$

However, at $i \neq i^{\bullet}$; $\sum_{j=0}^{m} \rho_{i,j+1} = \sum_{j=0}^{m} \rho_{ij}$; (i.e. $\rho_{i0} = \rho_{i,m+1}$) hence:

$$\sum_{j=0}^{m} \rho_{ij} = \sum_{j=0}^{m} \rho_{i-1,j}(1-k) + r_{i+1,j}k.$$

According to (11.55), this equation can be written as:

$$r_{i,j} = r_{i-1}(1-k) + r_{i+1}k. \tag{11.56}$$

Because it has been proved that the expression in Eq. (11.56) reflects the process equilibrium and stationarity, we can obtain principal dependencies of material balance without reverting to one-fold feeds and irrespective of time (Figure 11.1(a)). From the equilibrium conditions, we obtain:

$$\underbrace{r_i}_{\text{content in section } i} = \underbrace{r_i(1-k) + r_i k}_{\text{input to section } i}$$

and from the stationarity conditions:

$$\underbrace{r_i(1-k)+r_ik}_{\text{removal from section } i} = \underbrace{r_{i-1}(1-k)+r_{i+1}k}_{\text{input to section } i}$$

It follows that:

$$r_{i^*} = 1 = r_{i^*-1}(1-k) + r_{i^*+1}k \tag{11.57}$$

with the boundary conditions of the particles' escape from the apparatus:

$$\left.\begin{array}{r} r_1 = r_2k \\ r_z = r_{z-1}(1-k) \end{array}\right\}. \tag{11.58}$$

It can be noted that particles distribution by sections depends on the method of material feed. If we pass to fractional extraction of particles (F_f), this parameter will be invariable at various feeding regimes. By way of example, it can be demonstrated as follows.

Let us examine a stationary distribution in a continuous regime of feeding the initial material and periodic feed of a monofraction with the step $t = m + 1$. Let n single portions be fed into the channel in both cases with the intervals $1\cdot\Delta\tau$ and $(m+1)\Delta\tau$, respectively. During any ISA, $\rho_{1j}k$ particles of a narrow size class are always extracted from the upper part of the channel. In the first case, because particles content in all stages is stationary, the total amount of a fixed monofraction extracted upward is $P_1 = nr_1k$.

In the second case, at the feed of a single portion of particles, the output is expressed as follows:

$$\rho_2^1 = \rho_{10}k + \rho_{12}k + \ldots + \rho_{1m}k = k\sum_{j=0}^{m}\rho_{1j}.$$

According to Eq. (11.43):

$$\sum_{j=0}^{m}\rho_{1j} = r_1,$$

consequently,

$$P_2^1 = r_1k.$$

Respectively, at the feed of n unit portions, the extraction is:

$$P_2 = nr_1k.$$

Thus, in both cases, we arrive at the same fractional outputs. It can also be demonstrated for the case of a periodic feed regime with an arbitrary step $1 < t < m+1$. In order to pass to fractional extraction – the most important characteristic of the process – it is sufficient to examine a stationary regime of the apparatus operation with an interval $\Delta\tau$ between feeds.

In this regime, the extraction of a fixed monofraction into the fine product during an ISA amounts to $F_f(x) = r_1 k$. During n arbitrary ISAs, the value of fractional extraction is exactly the same, because in this case:

$$F_f(x) = \frac{nr_1 k}{n} = r_1 k.$$

To establish a functional dependence $F_f(x) = f(k;z;i;i^{\bullet})$, it is necessary to find a general expression of fractional particles content $r_1 = \varphi(k;z;i;i^{\bullet})$ described by a recurrent Eq. (11.56). To solve the latter, we apply the method of calculus of finite differences. We write Eq. (11.56) in the following form:

$$r_{i+1}k - r_i + r_{i-1}(1-k) = 0,$$

(i.e. we obtain a uniform linear finite-difference equation of the second order with constant coefficients). Its general solution is:

$$r_i = c_i\lambda_1^i + c_2\lambda_2^i \quad \text{(in case of } \lambda_1 \neq \lambda_2\text{)},$$

where λ_1 and λ_2 are roots of a corresponding characteristic equation:

$$\lambda^2 k - \lambda + 1 - k = 0. \tag{11.59}$$

Solving this equation, we obtain:

$$\lambda_1 = 1; \quad \lambda_2 = \frac{1-k}{k}$$

at $k \neq 0.5$ $\lambda_1 \neq \lambda_2$. Hence:

$$r_i = c_1 + c_2\left(\frac{1-k}{k}\right)^i. \tag{11.60}$$

To determine c_1 and c_2, we use boundary conditions from Eqs. (11.57) and (11.58) for $1 \leq i \leq i^{\bullet}$:

$$r_1 = r_2 k = \left[c_1 + c_2\left(\frac{1-k}{k}\right)^2\right]k = c_1 + c_2\frac{1-k}{k},$$

Expanding the left-hand side of the last equality, we obtain:

$$c_1 k + c_2 k\left(\frac{1-k}{k}\right)^2 = c_1 + c_2\frac{1-k}{k},$$

whence:

$$c_1 = \frac{c_2}{1-k}\left[k\left(\frac{1-k}{k}\right)^2 - \frac{1-k}{k}\right] = \frac{c_2}{1-k}\frac{1-k}{k}(1-k-1); \quad c_1 = -c_2. \tag{11.61}$$

To determine c_1, we use boundary condition (11.58).

At $i^{\bullet} \leq i \leq z$:

$$r_i = c_3 + c_4\left(\frac{1-k}{k}\right)^i. \tag{11.62}$$

Because $r_{i/i=z} = r_z = r_{z-1}(1-k)$, we derive from the expression in Eq. (11.60) as:

$$c_3 + c_4\left(\frac{1-k}{k}\right)^z = \left[c_3 + c_4\left(\frac{1-k}{k}\right)^{z-1}\right](1-k).$$

Solving this equation with respect to c_3, we obtain:

$$c_3 = -c_4\left(\frac{1-k}{k}\right)^{z+1}. \tag{11.63}$$

We substitute c_3 into Eq. (11.57):

$$r_{i^*} = 1 + \left[-c_2 + c_2\left(\frac{1-k}{k}\right)^{i^*-1}\right](1-k) + \left[-c_4\left(\frac{1-k}{k}\right)^{z+1} + c_4\left(\frac{1-k}{k}\right)^{i^*+1}\right]k. \tag{11.64}$$

According to the unambiguity condition, we also obtain:

$$r_{i^*/i\le i^*} = -c_2 + c_2\left(\frac{1-k}{k}\right)^{i^*} = r_{i^*/i\ge i^*} = -c_4\left(\frac{1-k}{k}\right)^{z+1} + c_4\left(\frac{1-k}{k}\right)^{i^*}. \tag{11.65}$$

Solving Eq. (11.65) with respect to c_2, we obtain:

$$c_2 = c_4\frac{\left(\frac{1-k}{k}\right)^{z+1-i^*} - 1}{\left(\frac{k}{1-k}\right)^{i^*} - 1}.$$

We substitute the result into Eq. (11.64):

$$1 + c_4\frac{\left(\frac{1-k}{k}\right)^{z+1-i^*} - 1}{\left(\frac{k}{1-k}\right)^{i^*} - 1}\left[-1 + \left(\frac{1-k}{k}\right)^{i^*-1}\right](1-k) + c_4\left[-\left(\frac{1-k}{k}\right)^{z+1} + \left(\frac{1-k}{k}\right)^{i^*+1}\right]k$$

$$= c_4\left[-\left(\frac{1-k}{k}\right)^{z+1} + \left(\frac{1-k}{k}\right)^{i^*}\right];$$

and hence:

$$c_4 = \frac{\left[1 - \left(\frac{k}{1-k}\right)^{i^*}\right]}{(1-2k)\left[1 - \left(\frac{1-k}{k}\right)^{z+1}\right]}.$$

Consequently:

$$c_2 = \frac{\left[1-\left(\frac{1-k}{k}\right)^{z+1-i^{\bullet}}\right]}{(1-2k)\left[1-\left(\frac{1-k}{k}\right)^{z+1}\right]}.$$

Substituting c_2 and c_4 into Eqs. (11.60) and (11.62) taking into account Eqs. (11.61) and (11.63), we obtain:

$$r_i = \frac{\left[1-\left(\frac{1-k}{k}\right)^{z+1-i^{\bullet}}\right]\left[1-\left(\frac{1-k}{k}\right)^{i}\right]}{\left[1-\left(\frac{1-k}{k}\right)^{z+1}\right](2k-1)} (1 \le i \le i^{\bullet}); \tag{11.66}$$

$$r_i = \frac{\left[1-\left(\frac{k}{1-k}\right)^{i^{\bullet}}\right]\left[\left(\frac{1-k}{k}\right)^{z+1}-\left(\frac{1-k}{k}\right)^{i}\right]}{\left[1-\left(\frac{1-k}{k}\right)^{z+1}\right](2k-1)} (i^{\bullet} \le i \le z). \tag{11.67}$$

In case of multiple roots of the characteristic Eq. (11.59) $\lambda_1 = \lambda_2 = 1$, at $k = 0.5$ the general solution is:

$$r_i = c_1 i\lambda^i + c_2\lambda^i = c_1 i + c_2. \tag{11.68}$$

Then, at $1 \le i \le i^{\bullet}$ $r_i = c_1 i + c_2$, and at $i^{\bullet} \le i \le z$ $r_i = c_3 i + c_4$.

Using the upper boundary condition $r_1 = r_2 k$, we obtain:

$$c_1 + c_2 = (c_1 2 + c_2)\frac{1}{2}$$

or

$$2c_1 + 2c_2 = 2c_1 + c_2.$$

Hence, $c_2 = 0$, and then:

$$r_{i/i \le i^{\bullet}} = c_1 i.$$

Using the lower boundary condition $r_z = r_{z-1}(1-k)$, we obtain:

$$c_3 z + c_4 = [c_3(z-1) + c_4]\frac{1}{2},$$

whence:

$$c_4 = -c_3(z+1).$$

Consequently:

$$r_{i/i\geq i^{\bullet}} = -c_2(z+1-i). \tag{11.69}$$

Using the single-valuedness condition, we obtain:

$$r_{i^{\bullet}/i\leq i^{\bullet}} = c_1 i^{\bullet} = r_{i^{\bullet}/i\geq i^{\bullet}} = -c_3(z+1-i^{\bullet}),$$

hence:

$$c_3 = -c_1\frac{i^{\bullet}}{(z+1-i^{\bullet})}.$$

Substituting the obtained expression into Eq. (11.69), we get:

$$r_{i/i\geq i^{\bullet}} = c_1\frac{i^{\bullet}}{(z+1-i^{\bullet})}(z+1-i). \tag{11.70}$$

We use the boundary condition of the feed:

$$r_{i^{\bullet}} = c_1 i^{\bullet} = c_1(i^{\bullet}-1)\frac{1}{2} + c_1\frac{i^{\bullet}}{z+1-i^{\bullet}}(z-i^{\bullet})\frac{1}{2} + 1.$$

Solving this equation with respect to c_1, we obtain:

$$c_1 = 2\frac{z+1-i^{\bullet}}{z+1}.$$

Substituting c_1 into Eqs. (11.68) and (11.70), we finally obtain:

$$r_i = \frac{(z+1-i^{\bullet})}{z+1}2i, \quad (1\leq i\leq i^{\bullet}); \tag{11.71}$$

$$r_i = \frac{2i^{\bullet}}{z+1}(z+1-i^{\bullet}), \quad (i^{\bullet}\leq i\leq z). \tag{11.72}$$

According to Eq. (11.66), fractional extraction into the fine product amounts to:

$$F(x)_{k\neq 0,5} = \frac{\left[1-\left(\frac{1-k}{k}\right)^{z+1-i^{\bullet}}\right]}{\left[1-\left(\frac{1-k}{k}\right)^{z+1}\right]}. \tag{11.73}$$

Using the expression in Eq. (11.71) for $i = 1$, we obtain:

$$F(x)_{k=0,5} = \frac{z+1-i^{\bullet}}{z+1}. \tag{11.74}$$

8. Optimisation of principal parameters of multi-stage separation

It is assumed that in critical regimes of two-phase flows a part of the solid phase moves along the flow, while another part settles against the flow.

Such a flow is extremely complicated and chaotic because of various random factors. Among them, we can mention mechanical interaction of particles of various sizes with each other and with channel walls, nonuniform concentration and velocity fields, multidirectional solid phase motion, presence of turbulent vortices and local irregularities, poly-fractional solid phase composition, irregular particles shape, etc.

Despite a huge number of papers dealing with the account for all these factors published in recent decades, the problem remains unsolved as yet.

Therefore, it is rather urgent to establish physically grounded similarity criteria for these flows.

Before developing a model of the process, let us examine the relation between the concentration and number of particles in real conditions in a narrow size class (e.g. with the average particle size of 30 μm). If the particles density is $\rho = 2600$ kg/m^3, then at the consumed concentration of $\mu = 2$ kg/m^3, volumetric concentration of the solid material in the air flow is insignificant and amounts to $7.7 \cdot 10^{-4}$ ($\beta = 2/2600$). The weight of one 30-μm particle is $0.3674 \cdot 10^{-11}$ kg. If they account for 10% only in the initial composition, then one cubic metre contains $n = 0.2/0.3674 \cdot 10^{-11} = 5.4 \cdot 10^9$ of them. It means that one cubic metre of air contains 5.4 milliard of these particles, which is comparable with the number of molecules in a rarefied gas.

To develop mathematical models of gaseous or liquid systems, Navier–Stokes equations are applied. It has been shown that after overcoming certain difficulties, these equations can be used for describing heterogeneous systems.

An equation for particles of one narrow class can be written with respect to vertical axis z as follows:

$$\frac{1}{\rho}\frac{\partial p}{\partial z} = P_z - \frac{dv_z}{dt} + c\nabla^2 v_z \qquad (11.75)$$

where ρ is the particles density, p denotes internal forces of the flow acting on particles accumulation in a unit mass; P_z denotes external forces acting on a unit particles mass; v_z is the velocity of particles of a narrow class in vertical direction; c is a proportionality coefficient; ∇^2 is a Laplacian (a sum of second partial derivatives of the value under the sign of this operator (v_z) with respect to coordinate axes ($x;y;z$). This parameter reflects tangential forces acting on a unit mass of particles.

In a general case, the velocity is a function of coordinates and time, and therefore its total differential equals:

$$dv_z = \frac{\partial v_z}{\partial t}dt + \frac{\partial v_z}{\partial z}dz + \frac{\partial v_z}{\partial x}dx + \frac{\partial v_z}{\partial y}dy.$$

We divide each summand in this expression by dt; keeping in mind that $dz/dt = v_z$; $dx/dt = v_x$; $dy/dt = v_y$, we can write:

$$\frac{dv_z}{dt} = \frac{\partial v_z}{\partial t} + v_z\frac{\partial v_z}{\partial z} + v_x\frac{\partial v_z}{\partial x} + v_y\frac{\partial v_z}{\partial y}. \qquad (11.76)$$

In our further derivation, we omit z for the sake of convenience.

It is impossible to obtain a general solution to Eq. (11.75) because of the lack of physical data. However, we can obtain similarity criteria for the process under study from Eq. (11.75).

Two physical phenomena are assumed to be similar if the homonymous properties in all similar points of geometrically similar channels differ by constant factors only. Mathematical descriptions of similar systems are identical.

We write Eq. (11.75) for two analogous points in similar systems in a stationary process taking into account Eq. (11.76):

$$\frac{1}{\rho_1}\frac{\partial p_1}{\partial z_1} = p_1 - \left(v_1\frac{\partial v_1}{\partial z_1} + \cdots\right) + c_1\left(\frac{\partial^2 v_1}{\partial z_1^2} + \cdots\right), \tag{11.77}$$

$$\frac{1}{\rho_2}\frac{\partial p_2}{\partial z_2} = p_2 - \left(v_2\frac{\partial v_2}{\partial z_2} + \cdots\right) + c_2\left(\frac{\partial^2 v_2}{\partial z_2^2} + \cdots\right). \tag{11.78}$$

The expressions in brackets are written in an abbreviated form. However, no information is lost, because the structure of all summands in brackets is the same.

By the definition of physical similarity, we can write:

$$\frac{\rho_1}{\rho_2} = m_\rho; \quad \frac{p_1}{p_2} = m_p; \quad \frac{P_1}{P_2} = m_m; \quad \frac{v_1}{v_2} = m_v; \quad \frac{z_1}{z_2} = m_l; \quad \frac{c_1}{c_2} = m_c.$$

The parameters m_ρ;m_p;m_m;m_v etc. are called similarity factors. Following these considerations, we can write:

$$\rho_1 = m_\rho \rho_2; \quad p m_p p_2; \quad P_1 = m_m P_2; \quad v_1 = m_v v_2; \quad z_1 = m_l z_2; \quad c_1 = m_c c_2.$$

Substitute the obtained expressions into Eq. (11.77):

$$\frac{m_p}{m_\rho m_l}\frac{1}{\rho_2}\frac{\partial p_2}{\partial z_2} = m_m P_2 - \frac{m_v^2}{m_l}\left(v_2\frac{\partial v_2}{\partial z_2} + \cdots\right) + \frac{m_c m_v}{m_l^2}\left(\frac{\partial^2 v_2}{\partial z_2^2} + \cdots\right). \tag{11.79}$$

This expression describes the system of Eq. (11.77) in notations of the system of Eq. (11.78).

By the definition of physical similarity, Eqs. (11.78) and (11.79) should be identical. It is possible in one case only – when the expressions composed from similarity factors are identical and can be taken out of brackets. Hence, we can conclude from Eq. (11.79) that:

$$\frac{m_p}{m_\rho m_l} = m_m = \frac{m_v^2}{m_l} = \frac{m_t m_v}{m_{lt}^2}. \tag{11.80}$$

Similarity criteria characterising the process can be obtained from Eq. (11.80). We are interested in a criterion reflecting physical fundamentals of a critical flow. Obviously, such criterion should be a measure of the ratio between gravity and inertia forces. Therefore, we examine only one equality of Eq. (11.80), namely:

$$m_m = \frac{m_v^2}{m_l}.$$

Let us disclose its content for a unit mass. By definition, we can write:

$$\frac{g_1}{g_2} = \frac{v_1^2 l_2}{v_2^2 l_1}.$$

Combining values with the same indices in different sides:

$$\frac{g_1 l_1}{v_1^2} = \frac{g_2 l_2}{v_2^2} = \frac{gl}{v^2} = Fr,$$

we obtain the Froude criterion (Froude number).

Earlier, the role of this criterion in the process under study has been revealed in a purely empirical way. This criterion allows obtaining universal separation curves when separating powders in turbulent regimes of the medium flow.

To find the parameters of separation curves affinization in all regimes, we make an attempt to solve Eq. (11.75) in a somewhat simplified form.

After overcoming certain difficulties, we can use equations derived in Chapter 2 for heterogeneous systems (i.e. particles motion in a flow):

$$-\frac{\partial n}{\partial t} = \frac{\partial (nv_x)}{\partial x} + \frac{\partial (nv_y)}{\partial y} + \frac{\partial (nv_z)}{\partial z} \quad (11.81)$$

or

$$\frac{\partial n}{\partial t} + \frac{\partial (nv_x)}{\partial x} + \frac{\partial (nv_y)}{\partial y} + \frac{\partial (nv_z)}{\partial z} = 0. \quad (11.82)$$

Clearly, integration of Eq. (11.82) with specified initial conditions (single-valuedness condition) leads to the conservation law.

Under certain conditions, Eq. (11.82) can be considerably simplified. First, a stationary process, in which all characteristics of a two-phase flow are time-independent, is of greatest interest. In such conditions:

$$\frac{\partial n}{\partial t} = 0.$$

Second, we can restrict ourselves with a quasi-one-dimensional model of transfer along the z axis, since the separation result depends only on the latter.

Reliable experimental data for determining transfer characteristics along the transverse axes (x,y) are unavailable as yet. Therefore, we assume that particles concentration changes only with height and remains unchanged over the cross-section at any level, the particles being uniformly distributed over this cross-section.

In this case:

$$\frac{\partial (nv_x)}{\partial x} = \frac{\partial (nv_y)}{\partial y} = 0, \quad (11.83)$$

and for Eq. (11.81) this condition is written as:

$$\frac{\partial n}{\partial x} = \frac{\partial n}{\partial y} = 0.$$

What remains from Eq. (11.82) is:

$$\frac{d(nv_z)}{dz} = 0,$$

and from Eq. (11.83):

$$v_z \frac{dn}{dz} = 0.$$

Mean flow velocity in a channel with a constant cross-section is constant; therefore, these two expressions are identical and can be written as:

$$\frac{dn}{dz} = 0. \tag{11.84}$$

This dependence reflects a certain condition of optimum distribution of solid particles in a flow with respect to the channel height.

By solving Eq. (11.84), we can find parameters for optimal organisation of the process of gravitational separation of particles in vertical channels.

Multidirectional motion of particles of one narrow size class takes place simultaneously at all levels over the channel height.

It is rather hard to develop a continuous mathematical model in this case. Therefore, we apply a somewhat different approach and develop a discrete model.

We mentally subdivide a vertical hollow channel into a finite number of segments (*N*), which we call stages. If the channel height is *H*, and the number of stages is *N*, the height of one stage is:

$$\Delta h = \frac{H}{N}.$$

The coordinate z can be written as $z = H - i\Delta h$. The derivative in Eq. (11.84) is:

$$\frac{\partial n}{\partial z} = \frac{\partial n}{\partial (H - i\Delta h)} = -\frac{\partial n}{\Delta h \partial i}.$$

We express the number of particles at a certain stage i as:

$$n_i = n_s r_i$$

where n_s is the quantity of particles in the initial material; r_i is the percentage of particles in the initial composition. It has been shown that in the site of material feed into the channel $r = 1$.

Because in the stationary regime material is not accumulated in the channel, at any stage a certain part of the material (*k*) rises, and another part $(1 - k)$ goes down. It is established that if all the sections are identical, and the flow velocity is constant, then the separation factor (*k*) is constant for all stages, for example:

$$k = const.$$

The separation factor is a function of the mean flow velocity, particles sizes and the diameter of the channel cross-section:

$$k = f(w; d; D).$$

As it has been agreed that the model under study is stationary and in equilibrium, it can be based on the material balance of a stage.

From the equilibrium condition:

$$r_i = r_i(1-k) + r_i k,$$

(i.e. the content of particles in stage *i* equals their transfer in both directions).

From the stationary state condition:

$$r_i(1-k) + r_i k = r_{i-1}(1-k) + r_{i+1}k,$$

(i.e. particles departure from the stage *i* equals their arrival at this stage).

Hence, we can obtain:

$$r_i = r_{i-1}(1-k) + r_{i+1}k \quad (i \neq i^*). \tag{11.85}$$

Boundary conditions for the initial material feed are:

$$r_{i^*} = 1 + r_{i^*-1}(1-k) + r_{i^*}k \tag{11.86}$$

and boundary conditions for particles departure from the channel are:

$$\begin{cases} r_1 = r_2 k \\ r_N = r_{N-1}(1-k) \end{cases}. \tag{11.87}$$

To determine the functional dependence:

$$r_i = r(k; N; i; i^*),$$

we make use of the recurrent Eq. (11.85). To solve this equation, we use the calculus of finite differences. We rewrite this equation in the form:

$$r_{i+1}k - r_i + r_{i-1}(1-k) = 0. \tag{11.88}$$

We obtain a system analogous to that obtained at the examination of a discrete-stationary system. Its solution comprises the following equations:

$$r_i = \frac{\left[1 - \left(\frac{1-k}{k}\right)^{N+1-i^*}\right]\left[1 - \left(\frac{1-k}{k}\right)^{i}\right]}{\left[1 - \left(\frac{1-k}{k}\right)^{N+1}(2k-1)\right]} \quad (1 \le i \le i^*), \tag{11.89}$$

$$r_i = \frac{\left[1 - \left(\frac{k}{1-k}\right)^{i^*}\right]\left[\left(\frac{1-k}{k}\right)^{N+1}\left(\frac{1-k}{k}\right)^{i}\right]}{\left[1 - \left(\frac{1-k}{k}\right)^{N+1}\right](2k-1)} \quad (i^* \le i \le N). \tag{11.90}$$

Fractional extraction of the size class under study upwards from the channel (into the fine product) at $i=1$ amounts to:

$$F_f(x) = r_1 k = \frac{\left[1 - \left(\frac{1-k}{k}\right)^{N+1-i^*}\right]}{\left[1 - \left(\frac{1-k}{k}\right)^{N+1}\right]}. \tag{11.91}$$

Fractional extraction into the coarse product at $i=N$ amounts to:

$$F_c(x) = z_z(1-k) = \frac{\left[\left(\frac{1-k}{k}\right)^{-i^*} - 1\right]\left(\frac{1-k}{k}\right)^{N+1}}{\left[1 - \left(\frac{1-k}{k}\right)^{N+1}\right]}. \tag{11.92}$$

The sum $F_f(x) + F_c(x) = 1$, which reflects the implication of the process (i.e. each narrow size class is completely distributed into both outlets).

Let us take a derivative of Eq. (11.89) with respect to z or i:

$$\frac{\partial r_i}{\partial i} = \frac{\left[1 - \left(\frac{1-k}{k}\right)^{N+1-i^*}\right]\left[\ln\left(\frac{1-k}{k}\right)\left(\frac{1-k}{k}\right)^{i}\right]}{\left[1 - \left(\frac{1-k}{k}\right)^{N+!}\right](2k-1)} = 0.$$

Because $\ln\left(\frac{1-k}{k}\right)\left(\frac{1-k}{k}\right)^{i} \neq 0$, consequently, $1 - \left(\frac{1-k}{k}\right)^{N+1-i^*} = 0$.

This equality is possible at any N and i^* values in one case only, at $1 - \left(\frac{1-k}{k}\right) = 1$ (i.e. at $k=0.5$). Another condition:

$$N + 1 - i^* = 0$$

does not make sense, because it leads to:

$$i^* = N + 1.$$

A derivative of Eq. (11.90) gives an analogous result. Thus, $k=0.5$ value corresponds to optimal separation within one stage. At $k=0.5$, for the entire channel:

$$F_f(x) = 0,5 = \frac{N+1-i^*}{N+1}.$$

Consequently:

$$i^* = \frac{N+1}{2}. \tag{11.93}$$

This means that in order to attain optimal separation, the initial material should be fed into the middle of the vertical channel.

It has been empirically established that to obtain high-quality separation, the initial material must not be fed from above or from below.

Substituting Eq. (11.93) into Eq. (11.91), we obtain:

$$F_f(x) = \frac{1}{1 + \left(\frac{1-k}{k}\right)^{\frac{N+1}{2}}}. \tag{11.94}$$

Reverting to Eq. (11.84), we proceed with the analysis of the obtained dependencies.

Let us take a derivative of $F_f(x)$ with respect to k:

$$\frac{dF_f(x)}{dk} = \left[\frac{\frac{N+1}{2}\left(\frac{1-k}{k}\right)^{\frac{N-1}{2}}\left(\frac{1}{k}\right)^2}{\left[1 + \left(\frac{1-k}{k}\right)^{\frac{N+1}{2}}\right]^2}\right]_{k=0.5} = \frac{N+1}{2}. \tag{11.95}$$

It is important to determine the limit fractional extraction of a narrow size that the class tends to with an increasing number of cascade stages and to find reasonable limits of the apparatus height. To do this, we write the Eq. (11.91) for a cascade apparatus in the general form:

$$F_f(x) = \frac{\left(\frac{k}{1-k}\right)^{z+1} - \left(\frac{k}{1-k}\right)^{i^*}}{\left(\frac{k}{1-k}\right)^{z+1} - 1}. \tag{11.96}$$

Let us examine possible alternatives:

1. $k > 0.5$ $\left(\frac{1-k}{k}\right) < 1$, in the limit, the denominator of Eq. (11.91) tends to unity.

Let us define the limit for this case:

$$\lim_{z\to\infty} F_f(x) = \lim_{z\to\infty}\left[1 - \left(\frac{1-k}{k}\right)^{z+1-i^*}\right] = 1 - \lim_{z\to\infty}\left(\frac{1-k}{k}\right)^{z+1-i^*}.$$

The final result depends on the ratio between z and i^* values:

a. at the material feed into the apparatus from above, $i^* = 1$,

$$\underset{z\to\infty}{F_f(x)} = 1 - \lim_{z\to\infty}\left(\frac{1-k}{k}\right)^z = 1.$$

In this case, all material comes out upward.

b. at the material feed from below, $i^* = z$,

$$F_f(x) = 1 - \lim_{z\to\infty}\left(\frac{1-k}{k}\right)^1 = 1 - \frac{1-k}{k} = \frac{2k-1}{k}.$$

This part of the material is extracted upward.

c. at the material feed in the middle part of the apparatus, $i^{\bullet} = (z+1)/2$,

$$\lim_{z\to\infty} F_f(x) = 1 - \lim_{z\to\infty}\left(\frac{1-k}{k}\right)^{\frac{z+1}{2}} = 1.$$

Finally, in this case, as well, all material is extracted into fine product.

2. If $k < 0.5$, then $\left(\frac{1-k}{k}\right) > 1$, and $\left(\frac{k}{1-k}\right) < 1$.
In this case, we can derive from (11.96) that:

$$\lim_{z\to\infty} F_f(x) = \lim_{z\to\infty}\frac{\left(\frac{k}{1-k}\right)^{z+1} - \left(\frac{k}{1-k}\right)^{i^{\bullet}}}{\left(\frac{k}{1-k}\right)^{z+1} - 1} = \left(\frac{k}{1-k}\right)^{i^{\bullet}}.$$

Let us examine possible alternatives of material feed into the apparatus for this case, as well:

a. feed from above, $i^{\bullet} = 1$, $\lim_{z\to\infty} F_f(x) = \frac{k}{1-k}$,

b. feed from below, $i^{\bullet} = z$, $\lim_{z\to\infty} F_f(x) = 0$.
In this case, all material goes out downward.

c. feed in the middle of the apparatus, $i^{\bullet} = (z+1)/2$, $\lim_{z\to\infty} F_f(x) = 0$.
In this case, too, all material goes out downward.

3. If $k = 0.5$, then $\lim_{z\to\infty} F_f(x) = \lim_{z\to\infty}\left[1 - \frac{i^{\bullet}}{z+1}\right]$.

In this case, as well, everything depends on z and $i^{\bullet}$ ratio:

a. at the feed from above, $i^{\bullet} = 1$, $F_f(x) = 1$.
All material rises upward;

b. at the feed from below, $i^{\bullet} = z$, $F_f(x) = 0$.
All material goes out downward.

c. at the feed in the middle of the apparatus, $i^{\bullet} = z + 1/2$, $\lim_{z\to\infty} F_f(x) = \lim_{z\to\infty}\left[1 - \frac{i^{\bullet}}{z+1}\right] = \frac{1}{2}$.

Thus, in this case only, irrespective of the number of stages, equal extraction of a narrow size class into both outlets is attained. As shown, this determines the optimality condition for the entire apparatus with respect to this boundary size.

To determine the separation optimality in a cascade in the general case at any $i^{\bullet}$ value, the value k is determined from the ratio:

$$\frac{1 - \left(\frac{1-k}{k}\right)^{z+1-i^{\bullet}}}{1 - \left(\frac{1-k}{k}\right)^{z+1}} = 0.5. \tag{11.97}$$

This clearly leads to the dependence of separation coefficient k on the quantity of stages and the site of material feed into the apparatus:

$$k = f(z; i^{\bullet}).$$

To attain optimality with respect to the boundary size value $x_{0.5}$, it is necessary to ensure the ascending flow velocity $w_{0.5}$ corresponding to the realisation of Eq. (11.97) relation (i.e. in the general case):

$$x_{0.5} = f(w_{0.};z;i^{\bullet}).$$

The obtained results can be illustrated by a particular example.

The conditions of separation optimality for a cascade separator consisting of 11 identical stages ($z = 11$) were determined for the entire apparatus and for one stage at a successive change in the material feed site from the upper ($i^{\bullet} = 1$) to the lower ($i^{\bullet} = 11$) stage. Principal results are summarized in the Table:

Dependence of fractional extraction of one stage and the entire apparatus on the site of the material feed into the apparatus under optimal conditions

Material Feed Stage, $i_{\bullet}$	1	2	3	4	5	6	7	8	9	10	11
Fractional extraction of the apparatus $F(x)$% at $k = 0.5$	91.6	83.3	75	66.7	58.3	50	41.7	33.3	25	16.7	8.3
Fractional extraction in one stage at $F(x) = 50\%$	0.33	0.42	0.45	0.47	0.485	0.5	0.515	0.535	0.55	0.585	0.667

Important conclusions follow:

1. Flow velocity equal to hovering velocity of boundary-size particles ensures the separation in one case only – at the material feed into the middle of the apparatus.
2. At any other feed site, hovering conditions do not ensure separation optimality. To attain optimality at the material feed above the middling stage, the flow velocity should be less than the hovering velocity, and at the feed below the middling stage, the flow velocity should exceed the hovering velocity.
3. All this points to the fact that the velocity hypothesis, which is a determining factor in today's theories of gravitational separation, is insufficient.
4. As follows from the Table, in the general case, separation optimality is determined not only by hydrodynamic properties of solid particles and the flow, but also, to a considerable extent, by the apparatus design and its boundary conditions.

Now let us revert to the relation in Eq. (11.95). Let us elucidate it using a graph of the function:

$$F_f(x) = f(k).$$

in Figure 11.5. This Figure shows an actual separation curve, and the path OABC shows ideal separation. The dependence of Eq. (11.95) shows the slope angle of a tangent at the point $F_f(x) = 0.5$. It is clear from the graph that the closer the curve to the ideal separation line, the higher the process efficiency. It means that the steeper the tangent under consideration, the more efficient the separation. It follows from Eq. (11.95) that the greater the number of stages, the higher the efficiency. However, such

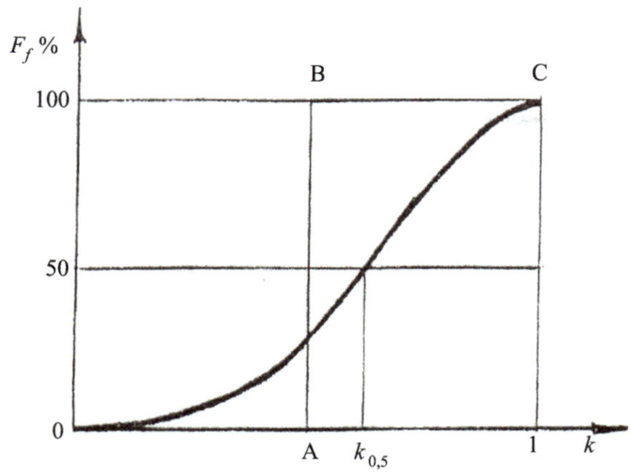

FIGURE 11.5 Fractional separation $F_f(x)$ dependence on the distribution parameter (k).

efficiency growth is limited. Therefore, the number of stages in one cascade should be a limited maximum by $z = 10 \div 11$. Further increase in the number of stages does not result in an appreciable growth of the effect.

CHAPTER

Application of the Obtained Results

12

CHAPTER OUTLINE

1. Invariants for estimating the perfection of the design of separation apparatuses 223
2. Method of theoretical estimation of vertical cascade separators 230
3. Experimental check of the adequacy of the method of cascade separators computation 232
4. Computation method of multirow combined separation cascade 235
5. Optimization of multirow classifier operation 239
6. Separator efficiency rank 242
7. Flow velocity ensuring optimal separation 243

1. Invariants for estimating the perfection of the design of separation apparatuses

Invariants derived and substantiated in the previous chapters constitute a strong basis for the progress in the theory and practice of separation of various materials.

At present, the main reason for the absence of an essential progress in separation technique is the absence of unbiased methods of evaluation and comparison of various separation facilities. This prevents an objective revelation of progressive tendencies in their designing and hinders the progress in this field of technology. Otherwise, it is hard to explain why many companies around the world have produced, until today, single-stage centrifugal separators, the main operation principles of which were patented as early as in the 19th century.

The first quality criteria for estimating the perfection of separation processes started appearing in the beginning of the 20th century, and a large number of these have been suggested. Disregarding the imperfection and narrowness of most of them, we can note that they are based only on the comparison of separated materials and final products composition. This is in no way connected with an attempt of unbiased estimation of the separating ability of a concrete apparatus. It is absolutely clear that different apparatuses have different separating abilities.

Due to such a situation, in most cases, the designing of new production processes connected with separation and optimization of working equipment has been based, until now, on previous experience in the use of similar equipment and on the intuition of the designer, and not on a scientific foundation.

Clarification and substantiation of the affinity of fractional separation curves cardinally changes the situation. This makes it possible to develop a quality criterion for an unbiased evaluation of the

Entropic Invariants of Two-Phase Flows. http://dx.doi.org/10.1016/B978-0-12-801458-5.00012-5

perfection of separation facilities design. The development of diverse cascade classifiers for various purposes has become possible in recent years owing to the establishment of such a criterion.

Separation curves of the $F_f(x) = f(w)$ type for a fixed narrow-sized class comprise all information about its mass distribution (Figure 12.1). It is clear, from these plots, that separation curves for a fixed velocity of a two-phase flow of $F_f(x) = \varphi(x)$ type can be obtained (Figure 4.3).

At that, universal curves obtained using transformation invariants, such as *Fr*, *B*, *Ba*, undoubtedly possess the highest community. As an example of affine transformation, it is shown how the curves presented in Figure 4.3 become one universal curve in the coordinate system $F_f(x) = f(Fr)$ (Figure 12.1). The invariants *B* and *Ba* are characterized by the greatest community.

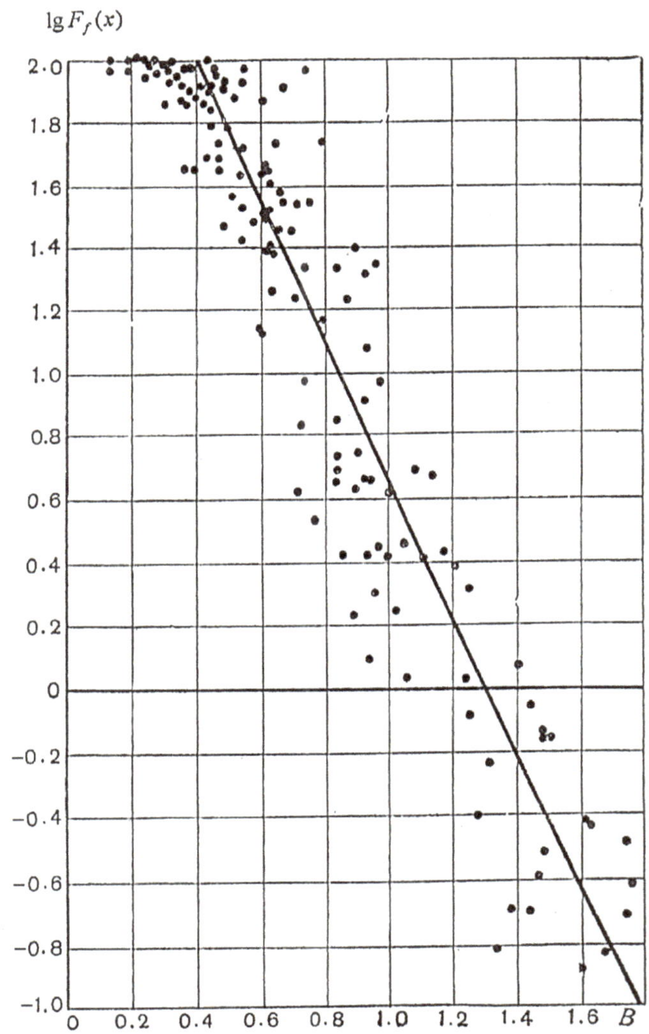

FIGURE 12.1 Semi-logarithmic dependence lg $Ff(x) = f(B)$ (by experimental data in Figure 13.4).

The assertion that separation curves represent the most important characteristic of the mass distribution process can be soundly attributed to universal curves. They also vary within the limits $0 \leq F_f(x) \leq 1$, are continuous, and are differentiable over the entire range of *Fr*; *B*; *Ba* values. They are characterized by a monotonous decreasing curve, for which

$$\frac{dF_f(x)}{dL} \leq 0$$

where L is one of invariants of (*Fr*; *B*; *Ba*) transformation. At that,

$$\left[\frac{dF_f(x)}{dL}\right]_{L=0} = \left[\frac{dF_f(x)}{dL}\right]_{L=\infty} = 0$$

This curve usually has two characteristic segments:

$$\frac{d^2F_f(x)}{dL^2} < 0 \quad \text{for the upper branch } (L < L_{50}),$$

$$\frac{d^2F_f(x)}{dL^2} > 0 \quad \text{for the lower branch } (L > L_{50})$$

where L_{50} is the invariant value at which $F_f(x) = 50\%$.

The lower branch extension over the range *L*, as a rule, significantly exceeds the upper branch extension. Therefore, these curves are well approximated in semi-logarithmic coordinates. Such approximation is valid within the range of fractional extraction variation from 0 to 97–98%. Experimental point deviations from it occur only in extraction regions close to 100%. Since the experiment accuracy, as a rule, does not exceed 4–5%, such deviations are within the limits of experimental accuracy. Surely, one can find a more complicated approximation for describing the entire range; however, in principle, it only complicates the expression without giving anything new, and therefore, we decide in favor of this one.

According to Figure 12.1, one can write for the universal curve

$$F_f(x) = A \cdot e^{-kL} \tag{12.1}$$

where $k = tg\alpha$, α is the angle of straight line slope to the abscissa axis.

Here, the kind of logarithm (common, natural, or binary) accepted is of no importance; they all differ by a constant. We can write

$$F_f(x) = A \cdot e^{-kL_{50}} = 50 \tag{12.2}$$

for a point where $F_f(x) = 50\%$, wherefrom we can determine the value *A*:

$$A = 50e^{kL_{50}} \tag{12.3}$$

Substituting Eq. (12.3) into Eq. (12.1), we obtain

$$F_f(x) = 50e^{k(L_{50}-L)} \tag{12.4}$$

The value $L_{50} = const$ for all narrow-sized classes, all regimes and all densities of the solid material and the medium in a specific separator, because the initial separation curves are affinely similar.

We derive Eq. (12.1) in the point L_{50} and write that

$$c = [F_f(x)]_{L_{50}} \cdot L_{50} \quad (12.5)$$

If we derive Eq. (12.1) and substitute $tg\alpha = k$, then Eq. (12.5) can be written as

$$c = AkL_{50}e^{-kL_{50}} \quad (12.6)$$

For the boundary condition, it follows from Eq. (12.1) that

$$50 = Ae^{-kL_{50}}$$

Taking this into account, the expression (12.6) acquires the form

$$c = \frac{kL_{50}}{2}$$

We introduce a parameter differing from the obtained one by a constant

$$\psi = kL_{50} \quad (12.7)$$

Note that the parameter ψ consists of a product of two values determining the position of the affine curve in the coordinate system. The value k estimates the steepness of the tangent to the universal curve at the point $F_f(x) = 50\%$, and L_{50} is the ordinate of the point where the entire set of technological parameters assures equal separation of all size classes into both outputs.

Obviously, it is practically impossible to draw an exact tangent at one point of an experimentally obtained curve at a natural spread of experimental points. However, as already noted, at a sufficient length of universal curve, its middle part is a straight line. It can be considered as the desired tangent, because it can be drawn more exactly, being based on a sufficient number of experimental points. In this respect, an approximation straight line obtained taking into account all experimental points is the most informative and exact. Apparently, this straight line is the most reliable for determining the parameter k, and, as for the L_{50} value, it is constant in any case. This approximation curve by itself also characterizes, in some way, the separating facility quality. The steeper the curve, i.e. the greater the parameter k, the higher separating ability of the apparatus. At an ideal separation, it should pass vertically through the point L_{50}. Note that the value of the parameter k proper is necessary, but not sufficient for such estimations of separators.

Let us try to understand this. Consider a universal curve in Figure 12.2. The most exact general characteristic of the separating ability of the apparatus can be derived from this curve, since it is unique for the apparatus under study. All experimental points obtained on this apparatus with changes in all conceivable technological parameters fit this curve. The apparatus imperfection is expressed in this graph by the hatched area. At an ideal separation, this area is zero; the smaller it is, the closer the separation to ideal one. It is rather complicated to use this area as a construction criterion, since it is difficult to determine it exactly from experimental data. Besides, to compare various apparatuses, one has to agree upon the scale of coordinate axes. We make an attempt to determine this area on the basis of the approximation that we have substantiated. This area can be represented as a sum of areas of two segments

$$S = S_1 + S_2$$

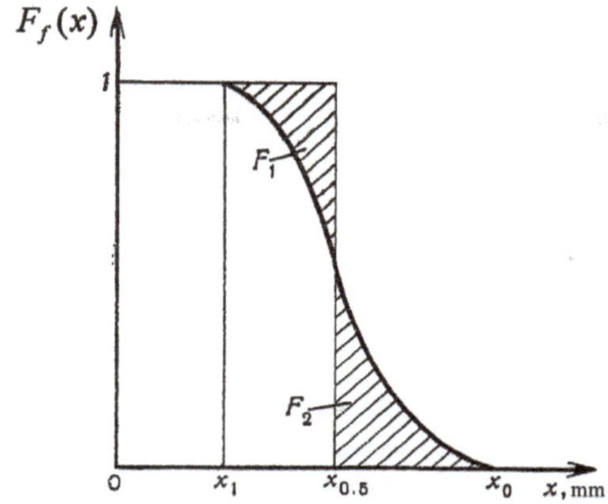

FIGURE 12.2 Determination of the separation completeness parameter.

The equation of universal curve can be written on the basis of Eq. (12.1):

$$F_f(x) = \ln A - kL \tag{12.8}$$

It follows from Figure 12.3 that

$$\left.\begin{aligned} A &= \psi + 3.9 \\ A &= z + 4.6 \end{aligned}\right\} \tag{12.9}$$

The first summand for the area S is determined as

$$S_1 = (L_{50} - L_{100}) - \int Le^{-kL}dL = (L_{50} - L_{100}) + \frac{e^z}{k}\left(e^{-kL_{50}} - e^{-kL_{100}}\right)$$

The second summand determined according to the same scheme is

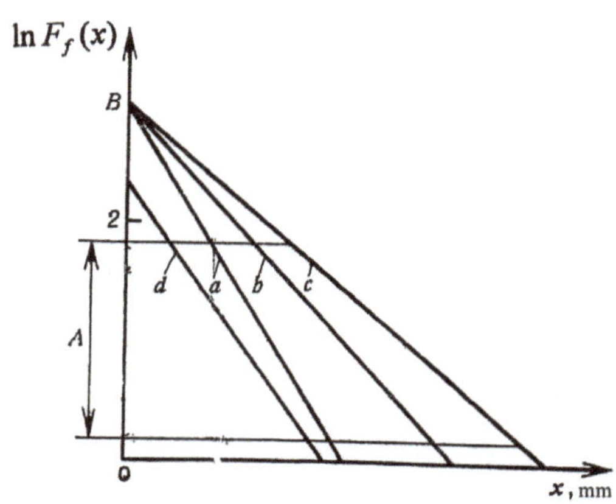

FIGURE 12.3 Principal relations for the approximating curve.

$$S_2 = \int_{L_{50}}^{L_{100}} Le^{-kL}dL = -\frac{e^z}{k}\left(e^{-kL_0} - e^{-kL_{50}}\right)$$

It follows from Figure 12.4 that

$$\left.\begin{aligned} L_{50}k &= \psi \\ z &= L_{100}k \\ \psi &= z + 0.7 \\ L_0k &= L + 4.6 \end{aligned}\right\} \qquad (12.10)$$

To attain unambiguous definiteness, we assume that the value L_0 corresponds to combined regimes that ensure 1% extraction of all narrow classes, i.e. $F_f(x) = 1\%$. It is impossible to determine zero extraction in logarithmic interpretation. Taking into account the dependence (12.10), the area S can be written as

$$S = S_1 + S_2 = (L_{50} - L_{100})\cdot 100 - \left[\frac{e^A}{k}\left(2e^{-\psi} - e^{kL_{100}} - e^{kz_0}\right)\right]$$

The expression in square brackets can be transformed:

$$\left[2e^{-z-0.7} - e^{-z} - e^{-z-4.6}\right] = e^{-z}\left[2\cdot 0.5 - 1 - e^{-4,6}\right] = e^{-z}\cdot\frac{1}{100}$$

Taking this into account, the total sum of areas is

$$S = (L_{50} - L_{100})\cdot 100 - \frac{e^A e^{-z}}{k}$$

We write the following ratio:

$$\frac{S}{L_{50}} = 100\left(1 - \frac{L_{100}}{L_{50}}\right) - \frac{e^A e^{-z}}{100kL_0} = 100\left[\left(1 - \frac{L_{100}}{L_{50}}\right) - 0.01\frac{1}{kL_{50}}\right]$$

Taking into account relations (12.10), we arrive at the dependence

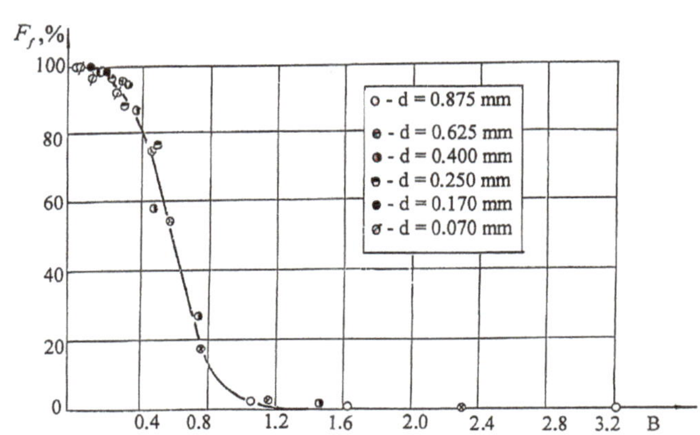

FIGURE 12.4 Ratio of estimated and experimental values of fractional separation of periclase.

$$\frac{S}{L_{50}} = 100\left[1 - \frac{z}{\psi} - \frac{0.01}{\psi}\right] = 100\left[\frac{\psi - z - 0.01}{\psi}\right] = 100\left[\frac{0.7 - 0.01}{\psi}\right]$$

Finally,

$$\frac{S}{L_{50}} = 100 \cdot \frac{0.69}{\psi} = \frac{69}{\psi}$$

whence we can obtain

$$\psi = 69\frac{L_{50}}{S} \tag{12.11}$$

The relation (12.11) allows us to draw some important conclusions.

First, for one apparatus, the parameter ψ is a constant. The relation (12.11) is the most general parameter reflecting the separation ability of a specific separator design, as it comprises the total area S and a constant parameter L_{50}.

Second, the magnitude of the obtained relation allows an understanding the physical sense of the previously formulated parameter ψ (12.7). It is an unambiguous criterion for estimating the quality of the separation ability of the apparatus.

Another point of interest is that, if the lines characterizing different separators start from the same point on the coordinate axis (Figure 12.3), they have the same ψ value. It means that their separation ability is the same. The higher the location of this point on the coordinate axis, the greater the parameter ψ. This parameter was defined as

$$\psi = kL_{50}$$

The value ψ can be determined even from the data of one experiment only:

$$\ln\frac{F_f(x)}{50} + kL = \psi \tag{12.12}$$

It is noteworthy that the relation (12.2) forms the basis of the process computation. The invariants F_r, B and B_a were determined on the basis of physics of flows and kinetic theory of two-phase flows, which include, taken together, all principal parameters of two-phase flows. Therefore, if we know the parameters ψ and k, we can compute all the results of initial material distribution and can predict the composition of separation products. Whenever necessary, it is possible, by varying structural elements, to select ψ and k so as to obtain required compositions of final separation products at a fixed initial feed. We show below how to determine the parameters ψ and k on the basis of approximating relation (Figure 12.5).

To determine the desired parameters, only two values of fractional extraction should be specified. It is desirable that one exceeded 50%, and another was below 50%, as shown in Figure 12.5. Consider first a triangle ABC, wherefrom we can find the parameter k as a tangent of the slope angle of the approximating curve to the abscissa axis:

$$tg\alpha = \frac{\ln F_1(x) - \ln F_2(x)}{L_2 - L_1} \tag{12.13}$$

$$k = tg\alpha$$

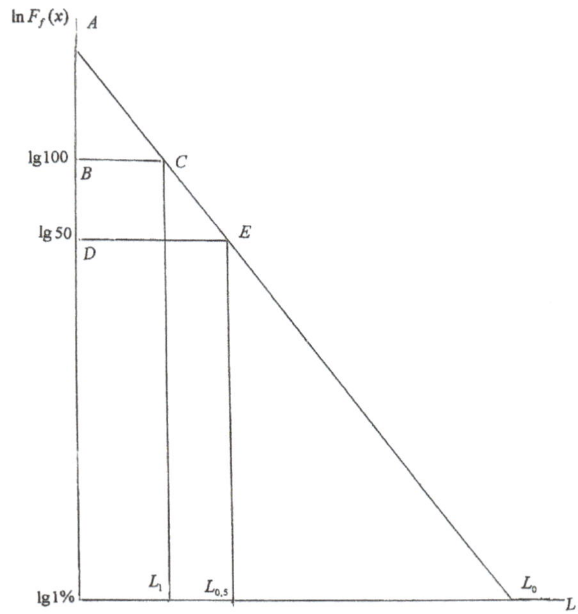

FIGURE 12.5 Estimation of separating capacity of a separator.

To determine L_{50}, we examine two similar triangles *ADE* and *ABC* and write

$$\frac{AD}{AB} = \frac{DE}{DC}$$

According to the graph, it can be written as follows:

$$\frac{\ln F_1(x) - \ln 50}{\ln F_1(x) - \ln F_2(x)} = \frac{L_{50} - L_1}{L_2 - L_1}$$

Hence,

$$L_{50} = L_1 + \frac{[\ln F_1(x) - \ln 50](L_2 - L_1)}{\ln F_1(x) - \ln F_2(x)} \tag{12.14}$$

All the parameters $F_1(x)$; $F_2(x)$; L_1; L_2 are known, and therefore, L_{50} can be determined unambiguously. This conclusion gives us a foundation for theoretical computation of separation parameters.

2. Method of theoretical estimation of vertical cascade separators

The set of theoretically derived relations allows us both to perform an exhaustive estimation of separating facility and to predict separation products compositions. Besides, optimal design of separators for obtaining powders of specified quality becomes possible. By way of example, we examine a complicated multistage cascade separator (not a simple one, distributing material in one stage).

Cascade character is usually determined by the design of internal facilities forming the cascade (Figure 4.11). Despite various structural settings, they all follow the same principle of multiple distribution of solid particles in two directions, owing to the impact of internal structural elements and carrier flow on the solid phase. These internal structural elements are arranged, as a rule, in a definite order.

Without going into an extremely complicated pattern of flow formation in real apparatuses, one can single out, at an oversimplified examination, the main characteristic connected with the separator design.

The velocity field of a continuum can be uniform or nonuniform. Hollow (equilibrium) apparatuses with a constant cross-section are the closest to the first case. A uniform velocity field causes the same regime of flow interaction with particles at any cross-section level over the separating column height and predetermines an extraction coefficient k or $F_f(x)$ invariable over the apparatus height. We assume that a cascade with $k = const$ for all stages is regular. A model of regular cascade should form the basis for calculating equilibrium apparatuses of circular, square, and rectangular cross-sections, as well as a hollow apparatus of zigzag type.

Conical apparatuses with expanding or converging cross-sections and a complicated configuration of walls are characterized by the distribution coefficient k varying over the height. At a known character of k variation, a model of a totally nonuniform cascade (with an arbitrary distribution coefficient) can serve as the basis for calculating such apparatuses. Shelf- and annular-type apparatuses, those with conical built-in elements, and some others are characterized by periodic narrowing and expansion of the fluid flow. This leads to a nonuniform velocity field and different intensities of the medium interaction with the material. Each stage of such apparatuses can be considered as consisting of two separating elements with the monofraction distribution coefficients k_1 and k_2. Therefore, a duplex model with alternating k_1 and k_2 should be the most suitable for predicting separation results in such conditions.

The intensity of medium interaction with particles depends not only on nonuniform velocity field over the apparatus height, but also on its cross-sectional nonuniformity. Thus, for an apparatus of circular cross-section, the effect of transverse nonuniformity of the flow is taken into account by the structural model. At that, it is assumed to be obvious that the extent of the apparatus cross-section filling with moving continuous phase is 100%. The situation is different in apparatuses of square and rectangular cross-section. In their corners, stagnant zones are formed, which reduce to zero the effect of medium on fine particle removal. Accepting a circumference inscribed into a square as a coarse estimate of the line of zero flow velocity in the apparatus of square cross-section, the extent of such cross-section filling with ascending flow is

$$c = \frac{F_0}{F} = \frac{\pi}{4}$$

Since the distribution coefficient is determined from structural model through the ratio of areas, a correction factor should be introduced at its computation for a square cross-section:

$$c = \frac{\pi}{4} \tag{12.15}$$

Hence,

$$k = \frac{\pi}{4} \cdot k_0 \tag{12.16}$$

where k_0 is the distribution coefficient for an apparatus of circular cross-section.

Accepting an inscribed ellipse as the line of zero flow velocity in the apparatus of rectangular cross-section, the expression (12.16) is also valid for determining k, since

$$c = \frac{F_e}{F} = \frac{\pi}{4}$$

The expression (12.16) for square and rectangular cross-sections is recommended as a first approximation only, since actual filling degrees are evidently higher.

Surely, the indices under consideration are mainly qualitative; however, they allow us to evaluate the acceptability degree of one model or another in each concrete case.

3. Experimental check of the adequacy of the method of cascade separators computation

In light of the above statement, an attempt at quantitative prediction of the results of a separation process was made on a number of apparatuses of different designs. Computation results are compared with experimental data presented in the respective figures. Figure 12.4 shows theoretical curves and experimental fractional extraction values at periclase classification (density $\rho = 3600$ kg/m^3) on an equilibrium apparatus of circular cross-section. The number of conventional stages $z = 9$, feed stage number $i^* = 6$. The consumed material concentration is $\mu = 1.5$ kg/m^2. The estimation was carried out using a regular cascade model according to the equation

$$F_f(x) = \frac{1 - \sigma^{z+1-i^*}}{1 - \sigma^{z+1}}$$

where $\sigma = \frac{1-k}{k}$ is the distribution index. The distribution coefficient k was determined according to a structural model by the formula

$$k = 1 - \sqrt{0.4 \cdot B}$$

where $B = \frac{\rho}{\rho_0} \cdot \frac{gd}{w^2}$ is a generalized classification parameter, $\rho_0 = 1.2$ kg/m^3 is the continuous medium density, $g = 9.81$ m/s^2 is the gravity acceleration, d is the average particle size, and w is the mean air velocity in the apparatus $\left(m/_s\right)$. Computed dependencies in this and the following graphs in this section are shown in the form of a continuous curve. Experimental data are shown by symbolic notations.

The average deviation of a theoretical curve from experimental points is +1.8–2.1%. Figure 12.6 shows the relation $F_f(x)$ at the classification of quartzite with the density $\rho_m = 2650$ kg/m^3 in an equilibrium apparatus of square cross-section. The number of conventional stages $z = 6$ and the feed stage number $i^* = 3$. Consumed material concentration $\mu = 2.0$ kg/m^3. The estimation was carried out using the relation

$$F_m = \frac{1 - \sigma^3}{1 - \sigma^7} \tag{12.17}$$

The distribution coefficient was determined with a correction for square cross-section according to Eq. (12.16)

$$k = \frac{\pi}{4}\left[1 - \sqrt{0.4 \cdot B}\right]$$

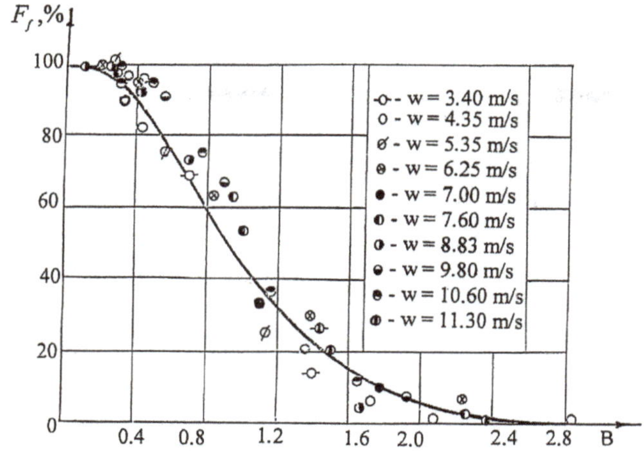

FIGURE 12.6 Ratio of estimated and experimental values of fractional separation of quartzite.

Maximal deviation of the theoretical curve from experimental points does not exceed 15%. Figure 12.7 shows the dependence $F_f(x)$ at the quartzite separation ($\rho_m = 2650$ kg/m^3) in a zigzag-type apparatus of a rectangular cross-section (Figure 12.8). The number of stages $z = 6; i^* = 3$. The consumed concentration $\mu = 2.0$ kg/m^3. The estimation was carried out using a regular cascade model:

$$F_f(x) = \frac{1 - \sigma^{z+1-i^*}}{1 - \sigma^{z+1}}$$

The distribution coefficient was determined with a correction for rectangular cross-section according to Eq. (12.10)

$$k = \frac{\pi}{4} \cdot \left[1 - \sqrt{0.4 \cdot B}\right]$$

The maximal deviation of the theoretical curve does not exceed 7%.

Figure 12.9 shows the results of the quartzite classification ($\rho_m = 2650$ kg/m^3) in a shelf cascade apparatus of a square cross-section. The number of stages in a cascade $z = 4$; $i^* = 1$. The consumed

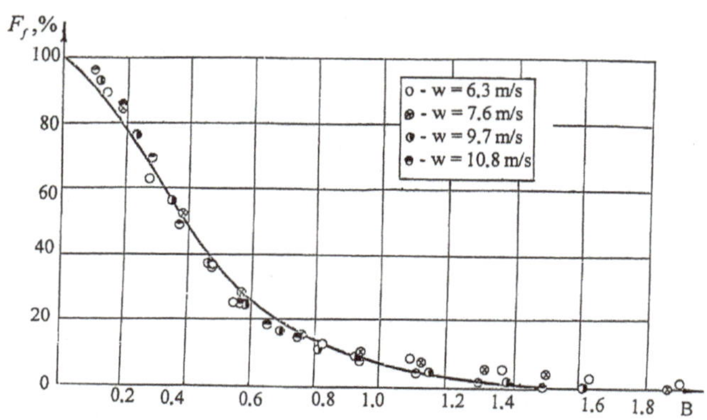

FIGURE 12.7 Ratio of estimated and experimental values of fractional separation on zigzag-type apparatus.

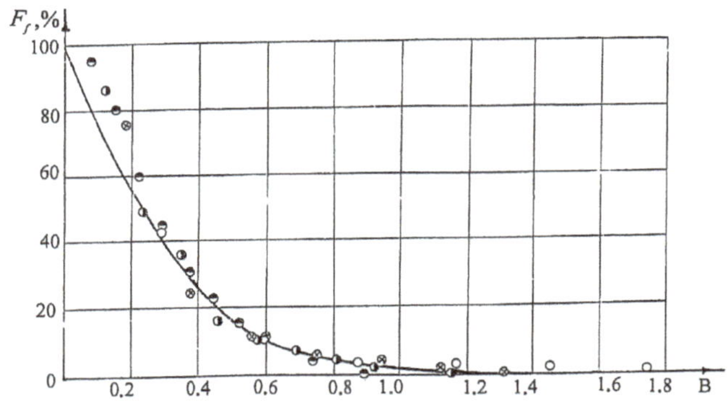

FIGURE 12.8 Results of separation on a shelf cascade of rectangular cross-section.

concentration $\mu = 2.0$ kg/m^3. The estimation of the entire cascade was carried out using a duplex $A \div B$ type cascade model:

$$F_m = \frac{1 - Q^{z-i^*} . \frac{1-k_2}{k_1}}{1 - Q^{z} . \frac{1-k_2}{k_1}} \tag{12.18}$$

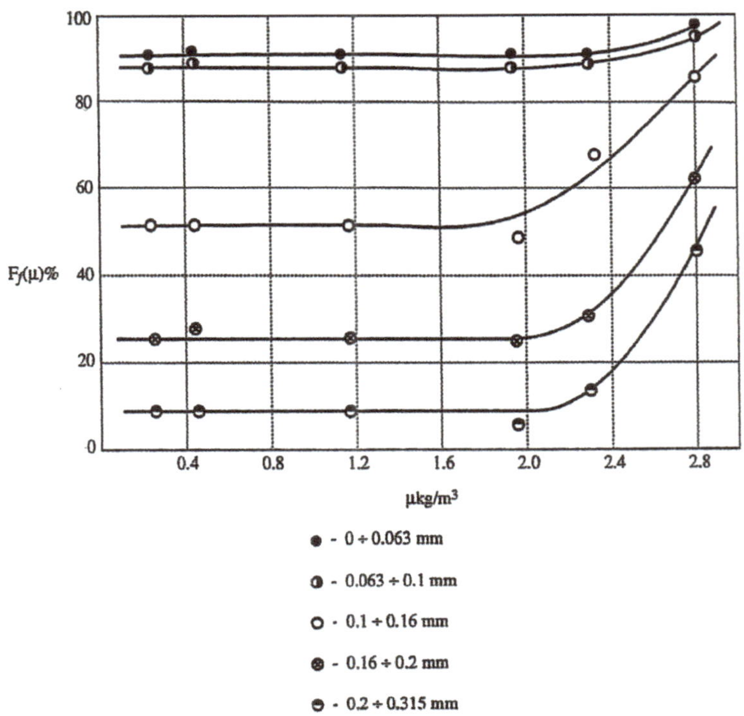

FIGURE 12.9 Fractional extraction dependence on the consumed solid phase concentration.

where $Q = \sigma_1 \cdot \sigma_2$ is a duplex parameter; $k_1 = \frac{\pi}{4}\left[1 - \frac{1}{2}\sqrt{0.4 \cdot B}\right]$ is a distribution coefficient for the element A (in the apparatus orifice) with a correction for rectangular cross-section; and $k_2 = \frac{\pi}{4}\left[1 - \sqrt{0.4 \cdot B}\right]$ is a distribution coefficient for the element of section B (in the overall cross-section of the apparatus) with a correction for square cross-section. Here the symbol A denotes the apparatus orifice where a cascade element is present, whereas B is the overall apparatus cross-section in the place of the absence of a cascade element. Maximal deviation does not exceed 10%.

We could give many more examples of such kind. On the whole, for all apparatuses under study, we can note a sufficient adequacy of estimated curves to experimental data.

4. Computation method of multirow combined separation cascade

The only drawback of the construction of vertical apparatuses is their limited productivity. They are usually designed for the initial material feed productivity of at most 10 ÷ 15 t/h. At high requirements to the quality of obtained products, the apparatuses productivity is limited by 5 ÷ 7 t/h. This drawback was overcome by the development of multirow combined cascade separators. By way of example, consider the results of industrial tests of a separator with the productivity of 40 t/h, the schematic diagram of which is shown in Figure 12.10. It was meant for obtaining a specified granulometric composition of fine-grained potassium chloride. The respective vendible product meant for exportation represented floto-concentrate dedusted by the boundaries of 0.1 and 0.2 mm with dust fraction content not exceeding 2%, while their content in the initial product was up to 20%.

The difficulty of the problem consisted in the fact that the available straight-through and circulation separators could not satisfy the requirements to the product quality. Studies of potassium chloride

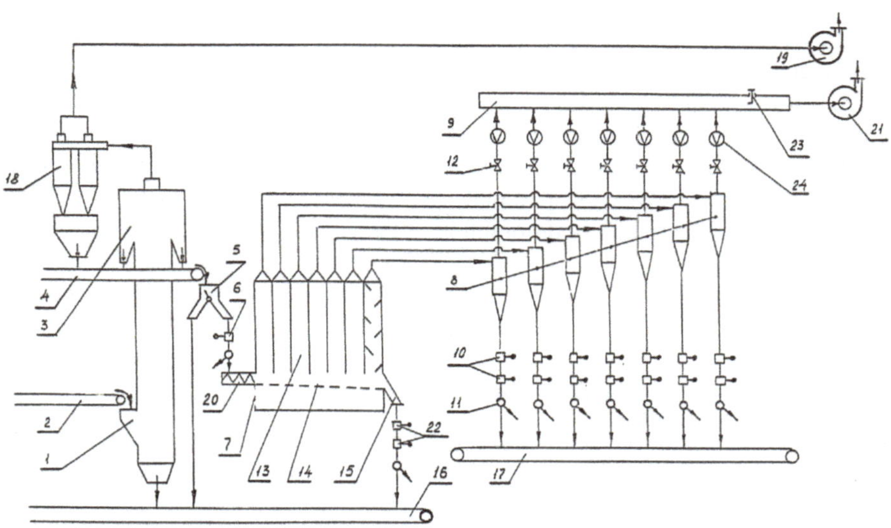

FIGURE 12.10 Scheme of a multirow separator.

dedusting performed on a centrifugal separator with the productivity up to 7 t/h have shown the possibility of decreasing the content of dust fractions in the coarse product down to 8% only. The possibility of solving this problem for the production of potassium fertilizers using a multirow separator was conjectured. To solve this problem, a multirow separator with the overall dimensions of 1.0 × 2.5 × 4 m and productivity of 40 t/h realizing the process of successive coarse product purification was developed and assembled.

A schematic diagram of the dedusting facility connection to the processing line is given in Figure 12.10. The apparatus consists of seven identical parallel separation columns. Each column (row) comprises six stages. In the lower part of the apparatus, a gas-distributing grate (a flat sheet with punched holes) is installed at a certain slope angle with respect to the horizon toward coarse product discharge.

The initial material was supplied from pipe-drier 1 through distribution facility 5 by screw feeder 20. The distribution facility allowed the variation of the material feed into the classifier from 5 to 45 t/h. The apparatus hermetization on the feeding side was realized by the screw feeder. The coarse product was discharged directly from the apparatus grate through a dual gate 22, ensuring the separation chamber hermetization on the side of the ready product discharge. The fine product was trapped in seven cyclones 8. The material was discharged from the cyclones onto the conveyer 17 through dual gates 10, ensuring the apparatus hermetization on the side of the dust product discharge. Air was supplied through window 7 due to rarefaction arising in the apparatus. A fan operated as a blowing machine. The facility operated under rarefaction, which excludes dust release into the environment. Before the discharge into the atmosphere, the air is subjected to two purification stages. The first stage is solid phase extraction in cyclones, and the second stage is particle trapping by foam apparatuses installed after the fan. To control air consumption in separating channels of the apparatus, gates 12 were installed on gas ducts between cyclones and common manifold 9, and diaphragms 24 for monitoring. Connecting pipes for material sampling were installed on the ducts of the initial material feed, coarse product discharge, and cyclone products.

The industrial testing procedure included the following:

1. A constant air consumption in the apparatus was established.
2. The initial material feed into the classifier and the apparatus operation mode were stabilized.
3. Samples for chemical and granulometric analyses were taken from the initial material and classification products using standard methods.
4. Material flows of outputs of classification products for a certain period of time were measured by cutting off dust fractions and coarse product.
5. During the entire test, the total air consumption in the apparatus and in individual separation columns was monitored.
6. Granulometric composition of the initial material and classification products was determined by screening the material on a set of sieves with meshes 0.8, 0.65, 0.4, 0.315, 0.2, 0.16, 0.1, and 0.063 mm.
7. Screening was performed mechanically by a standard method.

For a more extensive study of the capabilities of material classification in the apparatus, regimes of raw material dedusting by the four size boundaries 0.063, 0.1, 0.16, and 0.2 mm were investigated. In the process of the studies, fractionating regimes in the indicated size range and optimal productivity of the classifier were established, its separation ability was evaluated, and granulometric and chemical compositions of separation products were determined.

Since the self-similarity region of a multirow apparatus with respect to solid phase concentration in the gas flow had not been previously established in industrial conditions, we performed such research. Figure 12.9 shows fractional extraction dependence on the consumed concentration of the material at the air flow $Q = 15100\ m^3/h$. It follows from the graphs that, under industrial conditions as well as in laboratory experiments, self-similarity region with respect to consumed concentration lies within the limits $0 \div 2;\ 3\ kg/m^3$.

Figure 12.11 shows experimental dependencies of fractional extraction into fine product in individual columns of the apparatus for different monofractions. Graphical dependencies in Figure 12.11 allow estimation of the operation of a multirow apparatus using the principle of the one-row apparatus considered above. Here, continuous lines have been calculated using the model.

The model approximation of separator operation is based on the notion of uniformity, i.e. constancy of distribution coefficients of monofractions in identical sections of the apparatus and constancy of fractional extractions of individual columns. This statement was confirmed by these studies. In contrast to previously considered apparatuses, a multirow one has a special feature. It comprises two types of separating elements, i.e. a distributing grid and pouring shelves.

Pouring elements determine the operation of each section in two regimes. The aerodynamic regime in the narrow cross-section is characterized by the distribution coefficient of narrow class particles K_1, and in the total cross-section, respectively, by K_2. The mechanism of distribution coefficients formation was considered earlier. As a result, the following expressions are obtained for a shelf-type column of square or rectangular cross-section:

$$K_1 = 0.8678\left(1 - 0.5\sqrt{0.4 \cdot B}\right)$$

$$K_2 = 0.8678\left(1 - \sqrt{0.4 \cdot B}\right)$$

The coefficient 0.8678 is due to the use of a third-power equation to characterize the extent of cross-section area filling with the continuous medium.

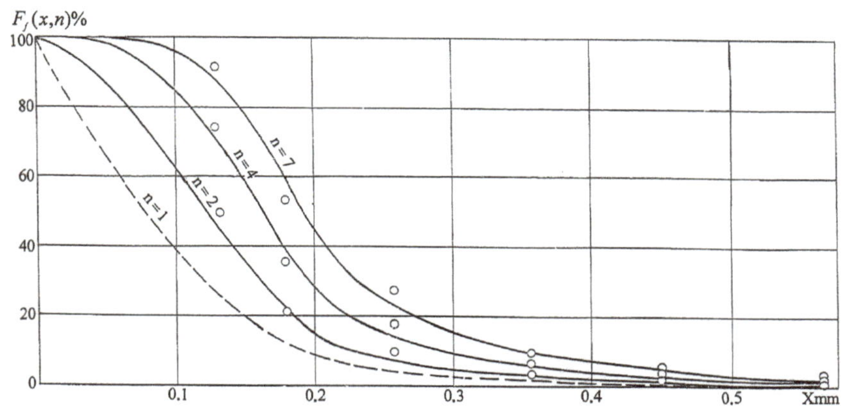

FIGURE 12.11 Fractional extraction dependence on the number of cleaning rows.

Fractional extraction of a single cascade column without a grid of $A \div B$ type at the initial flow feed into the element B is determined, according to a duplex cascade model, by Eq. (12.17):

$$F_{0_2} = \frac{1 - Q^{z-i^*} \cdot \frac{1-k_2}{k_1}}{1 - Q^z \cdot \frac{1-k_2}{k_1}}$$

For a specific multirow apparatus design, $i^* = z$, and hence

$$F_{0_2} = \frac{1 - \frac{1-k_2}{k_1}}{1 - Q^z \cdot \frac{1-k_2}{k_1}}$$

The distributing grid, which is a separating unit working in combination with a cascade column, is characterized by a coefficient of fixed monofraction transfer from the grid to the lower section of the column λ.

A semi-empirical relation for the transfer coefficient λ is found, which connects it with the generalized classification parameter:

$$B = 3 \cdot \frac{\sqrt{1 - \sqrt{\lambda}}}{1 + 2\sqrt{\lambda}} \tag{12.19}$$

Distributing grid combination with the column realizes consecutive operation with partial monofraction recirculation from the column behind the grid (Figure 12.12). The share of active recovery is about half of the flow of particles of a fixed narrow-sized class of coarse product, because the recovery is related to half of the grid area under the next column (Figure 12.12). According to this structural scheme of the grid connection with the column, fractional extraction achieved by their joint operation is expressed by the relation

$$F_0 = \frac{\lambda F_{0_2}}{1 - \frac{1}{2}\lambda(1 - F_{0_2})} \tag{12.20}$$

A general expression for fractional extraction of particles into the fine product for the entire combined separating cascade has the form

$$F = 1 - (1 - F_0)^n \tag{12.21}$$

Graphical dependence $\lambda(B)$ according to Eq. (12.19) is shown in Figure 12.13.

Thus, using the derived relationships, we obtain a complete analytical estimation of a multirow separating cascade.

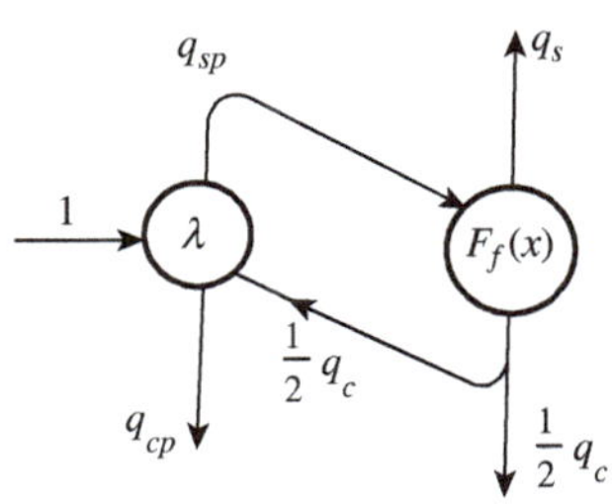

FIGURE 12.12 Analytical model of joint operation of a column and a grate.

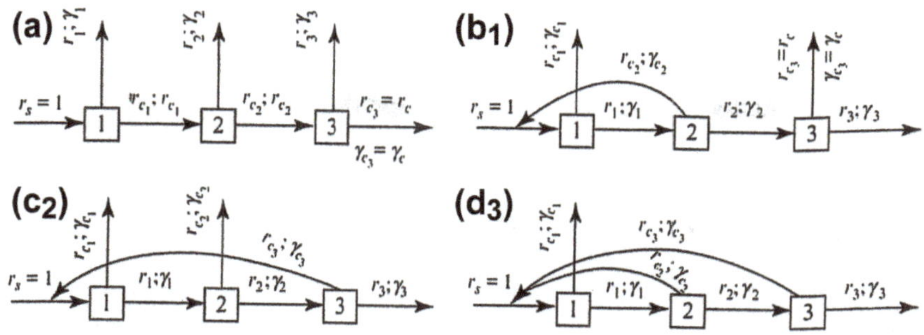

FIGURE 12.13 Multirow separator.

(a) with an open cycle; (b) with a recycle of the second stage; (c) with a recycle of the third stage; and (d) with a recycle of the second and third stages.

5. Optimization of multirow classifier operation

A multirow cascade classifier comprising 12 rows of cascade separation with the productivity of 60–70 t/h was developed (Figure 12.1). The apparatus is meant for fractionating potassium chloride containing at least 80% of material with the size exceeding 0.1 mm. For our analysis, we subdivide it into three sections of four rows each.

According to technological conditions, dust fraction (−0.1 mm) content in a dedusted product should not exceed 2%. To calculate the fractionation scheme, we apply relations obtained in previous sections. The calculation is performed for optimal flow velocity that, according to industrial tests, amounts to 4.1 m/s. First, we examine the apparatus operation without recirculation. Experimental separation curves depending on the number of rows are shown in Figure 12.11.

Products outputs from the apparatus sections are determined using the following formulas:

Section 1

$$\gamma_1 = \sum r_s F_4(x);$$

Section 2

$$\gamma_2 = \sum r_s [F_8(x) - F_4(x)];$$

Section 3

$$\gamma_3 = \sum r_s [F_{12}(x) - F_8(x)].$$

Product composition is determined as follows:

Section 1

$$r_1 = \frac{r_s F_4(x)}{\gamma_1};$$

Section 2

$$r_2 = \frac{r_s[F_8(x) - F_4(x)]}{\gamma_2};$$

Section 3

$$r_3 = \frac{r_s[F_{12}(x) - F_8(x)]}{\gamma_3}.$$

Contamination of coarse product calculated by this scheme equals 0.4%. For this scheme of the process organization, in which all fine product outputs are united to form the dust fraction, the following process index is obtained:

$$\gamma_{\text{int}} = 64.3\%; \quad D_{-0,1} = 0.4.$$

In a dedusted product $D_{-0.1} = 0.4\%$, which satisfies technical conditions. However, the respective dedusted product output is rather low (64.3%), and, besides, a significant part of the +0.1-mm class is lost in the fine product (up to 15.62% of the initial one). Here the +0.1-mm class losses in sections 2 and 3 with their minor outputs (7.08% and 4.28%, respectively) amount to 8.71% of the initial content, i.e. to half of the total losses.

Apparently, it is expedient not to mix the fine products of Sections 2 and 3 with the dust product of Section 1, but to classify them additionally in the same apparatus.

We define an optimal scheme of fractionation with recirculation. We calculate fractional extraction of narrow-sized classes into the fine product for the case corresponding to Figure 12.13b.

The following notations are accepted: $r_{f_1}; r_{f_2}; r_{f_3}$ are the contents of a certain narrow-sized class in the fine products of Sections 1, 2, and 3, respectively; $r_{c_1}; r_{c_2}; r_{c_3}$ are the contents of the same-sized class in the coarse products of Sections 1, 2 and 3, respectively.

In our calculation, we use the following formulas:

$$F_f(x) = \frac{r_f}{r_s}\gamma_f$$

or

$$\begin{aligned}
F_f(x) &= \frac{r_{f_1}}{r_s}\gamma_{f_1} + \frac{r_{f_3}}{r_s}\gamma_{f_3};\\
F_{f_1} &= \frac{r_{f_1}}{r_s + r_{f_2}\gamma_{f_2}}\gamma_{f_1};\\
F_{f_2} &= \frac{r_{f_2}}{r_{c_1}}\frac{\gamma_{f_2}}{\gamma_{c_1}};\\
F_{f_3} &= \frac{r_{f_3}}{r_{c_2}}\frac{\gamma_{f_3}}{\gamma_{c_2}};\\
r_s &= r_{f_2}\gamma_{f_2} = r_{f_1}\gamma_{f_1} + r_{c_1}\gamma_{c_1}\\
r_{c_1}\gamma_{c_1} &= r_{f_2}\gamma_{f_2} + r_{c_2}\gamma_{c_2}\\
r_{c_2}\gamma_{c_2} &= r_{f_3}\gamma_{f_3} + r_{c_3}\gamma_{c_3}\\
r_s &= r_{f_1}\gamma_{f_1} + r_{f_3}\gamma_{f_3} + r_{c_3}\gamma_{c_3}
\end{aligned} \tag{12.22}$$

where $F_f(x)$ is a fractional extraction of a fixed fraction into the fine product for the entire apparatus; $F_{f_1}; F_{f_2}; F_{f_3}$ – fractional extraction of a fixed fraction into the fine product for sections 1, 2, and 3, respectively.

For the case under study,

$$F_f = \frac{r_{f_1}}{r_s}\gamma_{f_1} + \frac{r_{f_3}}{r_s}\gamma_{f_3} \tag{12.23}$$

Hence,

$$F_f = F_{f_1}\left[1 + F_{f_2}\frac{1}{(1-F_{f_2})(1-F_{f_3})}\frac{r_{c_3}}{r_s}\gamma_c\right] + F_{f_3}\frac{1}{1-F_{f_3}}\frac{r_{c_3}}{r_s}\gamma_3 \tag{12.24}$$

We finally obtain

$$F_f = \frac{F_{f_1} + F_{f_3} - F_{f_1}F_{f_2} - F_{f_2}F_{f_3} + F_{f_1}F_{f_2}}{1 - F_{f_1} + F_{f_1}F_{f_2}} \tag{12.25}$$

Under the condition that each section has four separation columns:

$$F_{f_1} = F_{f_2} = F_{f_3} = \left[1 - (1 - F_{f_1})^4\right] \tag{12.26}$$

Transforming the previous dependence, we obtain:

$$F_f = \frac{1 - F_{c_1}^1 + F_{c_1}^8 F_{c_1}^{12}}{1 - F_{c_1}^4 + F_{c_1}^8} \tag{12.27}$$

where $F_{c_1} = 1 - F_{f_1}$ is the extraction of a certain narrow-sized class into the coarse product in the first separation column. Hence,

$$F_c = \frac{F_{c_1}^{12}}{1 - F_{c_1}^4 + F_{c_1}^8} \tag{12.28}$$

For the case of Figure 12.13c

$$F_c = \frac{F_{c_1}^{12}}{1 - F_{c_1}^8 + F_{c_1}^{12}} \tag{12.29}$$

For the case of Figure 12.13d

$$F_c = \frac{F_{c_1}^{12}}{1 - F_c^4 - F_c^{12}} \tag{12.30}$$

Thus, depending on the scheme of fine-product recirculation, we can select an optimal scheme of the process organization.

Using the presented method, the results for the three cases are estimated and summarized for comparison in Table 12.1.

Thus, the variant III (a scheme with recirculation of the product of Sections 2 and 3) is the most efficient for potassium chloride dedusting. It allows an increase in the ready product output by 5% and

Table 12.1 Results of Fractionation Scheme Estimation

Variant	γ_c, %	$D_c^{(-0,1)}$, %	$\gamma_{n.n}$, %	$R_{n.n}^{(+0,1)}$	$R_{(s)}^{(+0,1)}$
I	66.610	0.434	33.390	40.069	13.3
II	66.001	0.374	33.999	41.011	13.94
III	66.682	0.449	30.318	34.087	10.33

a decrease in target class removal into the dust fraction down to 10.33% (as compared with 15.62% in the open cycle).

The apparatus productivity increase by 11% with respect to the initial feed does not change the classifier operation regime in the region of self-similarity.

This example clearly demonstrates the capabilities of the combined principle of cascade separation organization.

6. Separator efficiency rank

The parameter of separating capacity of the apparatus ψ suggested in this chapter has a number of advantages. However, it is not convenient enough, since it varies from 0 to ∞. This makes it difficult to understand how perfect the apparatus design is on the basis of ψ value. Therefore it is expedient to define a criterion varying within a habitual 100% scale on the basis of this parameter.

Following Figure 12.5, we can write

$$\frac{\psi}{L_{50}} = \frac{\psi - 0.7}{L_{100}}$$

whence

$$L_{100} = L_{50}\left(1 - \frac{0.7}{\psi}\right)$$

Introducing a notation

$$E = \left(1 - \frac{0.7}{\psi}\right) \cdot 100\% \tag{12.31}$$

we can readily show that the parameter E is monotonically connected with the parameter ψ, i.e. the higher separating ability, the greater the expression (12.31). We examine this relation in more detail.

First, note that the coefficient 0.7 is a minimum possible value of the parameter ψ. It means that we deal with an apparatus in which $F_f(x) = 100\%$ for any narrow-sized class is possible only for $L = 0$. In this case, flow velocity w should tend to infinity, or else the particle size should tend to zero, which is impossible to imagine.

Second, this parameter possesses all features required for objective reflection of separation ability quality. It acquires maximal value $E = 100\%$ for an ideal separator ($\psi \rightarrow \infty$). For the worst separator ($\psi = 0.7$), its value is $E = 0$. Thus, it changes monotonically from 0% to 100%. At the same time, it is independent of the initial feed composition, boundary size of separation and of the totality of regime

Table 12.2 Dependence of Separator Rank E% on the Apparatus Heights

	Number of stages						
	1	2	4	6	8	12	14
Cascade	69	69.8	70.2	71	71.2	71.6	71.7
Zigzag	–	57.6	62	65	69.3	–	–
Equilibrium	50.4	57	62.5	63.7	65.6	–	–

factors of the process. Since $\psi = kL_{50}$, using this index, we can find all optimal separation regimes, because

$$\psi = \frac{0.7}{1 - E} \tag{12.32}$$

It is of interest to determine this parameter for some of examined separators in order to obtain deeper insight into its effect. Table 12.2 shows the dependence of this parameter on the number of stages of various types of cascade separators.

It follows from this table that the effect of height is unambiguous. Separation quality grows with increasing apparatus height. It reliably reflects the parameter of separator rank.

It is of interest to check the functioning of apparatuses with various cross-section areas. It can be shown by experimental data for cascade separators with the same number of stages having different values for the equivalent diameter, the smallest being taken as unity.

Here we observe an unambiguous trend—with increasing separator cross-section, its separating ability decreases. Awareness of this fact led to the development of a multirow separator with each element having a limited cross-section, while in sum they ensure a high productivity at a high separation quality.

7. Flow velocity ensuring optimal separation

As follows from Eq. (12.4), a general dependence of fractional extraction of a narrow-sized class can be written as

$$F_f(x) = 50e^{k(L-L_0)} \tag{12.33}$$

where L_0 is the invariant value at which any narrow size class x_0 becomes a boundary one, i.e. is extracted by 50%, and L can be any of the three derived invariants (Table 12.3).

Table 12.3 Separator Rank Dependence on the Scale Factor

Scale factor, D_ε	1	1.4	1.75	2.0	2.45	5.06
Separator rank, E%	80.3	72.4	70.2	61.8	51	20

We take one of them for our purposes, for example, the parameter B. Equation (12.31) can be rewritten for it as follows:

$$F_f(x) = 50e^{-\frac{kg}{w^2}\frac{(\rho-\rho_0)}{\rho_0}(x-x_0)} \tag{12.34}$$

This dependence allows the establishment of a connection between the extraction of various classes with respect to the boundary size; for $x = x_0$ the value $F_f(x) = 50\%$; for $x < x_0$ $F_f(x) < 0$, and for $x < x_0$ $F_f(x) > 50\%$. Maximal narrow class extraction is 100%. Therefore, according to Eq. (12.34), we can write:

$$50e^{-\frac{kg}{w^2}\frac{(\rho-\rho_0)}{\rho_0}(x-x_0)} = 100$$

Hence,

$$-\frac{kg}{w^2}\frac{(\rho-\rho_0)}{\rho_0}(x-x_0) = \ln 2$$

Minimal velocity of 100% extraction of a certain size class is

$$w_{\min} = \sqrt{-\frac{kg(\rho-\beta)(x-x_0)}{\rho_0 \ln 2}} \tag{12.35}$$

Maximal size of particles extracted by 100% at a fixed flow velocity is determined by the expression

$$x_{\max} = x_0 - \frac{w^2 \ln 2\rho_0}{gk(\rho-\rho_0)} \tag{12.36}$$

Previously it was agreed to assume 1% extraction as the minimal one. Taking into account experimental error, at a lower extraction, the logarithm of this value introduces uncertainty into the location of experimental points on the graph. For this extraction value, we can write

$$50e^{-\frac{kg(\rho-\rho_0)}{w^2\rho_0}(x-x_0)} = 1$$

It follows that

$$-\frac{kg(\rho-\rho_0)}{w^2\rho}(x-x_0) = \ln 0.02$$

Hence, maximal velocity at which a narrow class is extracted by 1% is

$$w_{\max} = \sqrt{-\frac{kg(\rho-\rho_0)}{\ln 0.02\rho_0}(x-x_0)} \tag{12.37}$$

At that, the minimal particle size is

$$x_{\min} = x_0 - \frac{w^2 \ln 0.02\rho_0}{gk(\rho-\rho_0)} \tag{12.38}$$

Thus, particles of a certain range [see Eq. (12.37)] that can be represented as

$$x_0 + \frac{3.9w^2\rho_0}{gk(\rho - \rho_0)} \le x_i \le x_0 - \frac{0.7w^2\rho_0}{gk(\rho - \rho_0)} \tag{12.39}$$

experience separation in a flow. Hence, it is clear that, first, an ideal separation ($x_i = x_0$) in a flow is impossible. We can only set a problem of the convergence of interval boundaries due to increasing parameter k connected with separating ability of the apparatus. Second, such analysis shows that at particles hovering velocities exceeding flow velocity ($w_0 > w$), these particles have a chance to be withdrawn into the fine product, but at the same time, not all particles with $w_0 < w$ are extracted into the fine product.

We can also conclude that particles of different-sized classes extracted into the fine product on the same apparatus in the same way have the same relations between the flow velocity and particle size due to the affinity of the invariant value $L = const$.

First, we examine an optimal regime. Under such conditions, we can write the following for particles of all classes:

$$w_{50} = \sqrt{\frac{gd}{B_{50}} \frac{(\rho - \rho_0)}{\rho_0}}$$

Since, on the other hand,

$$w_{50} = c\sqrt{\frac{4dg(\rho - \rho_0)}{3\lambda\rho_0}}$$

where c is a certain structural coefficient, $\frac{1}{\sqrt{B_{50}}} = c\sqrt{\frac{4}{3\lambda}}$. whence

$$B_{50} = \frac{3\lambda}{4c^2} \tag{12.40}$$

In this expression, λ is a resistance coefficient for a unit particle. It has a constant value in the turbulent flow region. In these conditions $B_{50} = const$, the structural coefficient c has a constant value at a fixed extraction of narrow-sized classes.

The total resistance of two-phase flow is composed of respective resistances of each size class and depends on the content of these classes in a mixture. Here we are already close to the understanding of the structure of resistance formation in a two-phase flow in the regimes of solid phase distribution. Optimal value of the invariant B can be written as

$$B_{50} = \frac{\psi}{k} \tag{12.41}$$

Taking this into account, we obtain

$$\frac{3\lambda}{4c^2} = \frac{\psi}{k} \tag{12.42}$$

Since the structural coefficient appears in invariants through the estimated flow velocity, it is rigidly connected with fractional separation value $F_f(x)$ for particles of each narrow-sized class. This allows us to conclude that the dynamic resistance coefficient of the process depends not only on the apparatus construction c and particle resistance coefficient λ, but also on the fractional separation value $F_f(x)$. In

Table 12.4 Dependence of Separation Indices on the Place of Material Feed Into a Cascade Apparatus

Indices	Material feed stage										
	1	2	3	4	5	6	7	8	9	10	11
Fractional extraction $F_f(x)$, % for $k=50\%$	91.6	83.3	75	66.7	58.3	50	41.7	33.3	25	16.7	8.3
Fractional extraction for one stage $k\%$ for $F_f(x)=50$	33	42	45	47	48.5	50	51.5	53.5	55	58.5	66.7

this connection, it becomes necessary to introduce some corrections into the existing theoretical opinions on the mechanism of the process under study. In the gravitational concentration theory, the main attention is paid to the establishment of the final precipitation velocity (or hovering velocity) of particles of boundary size depending on solid material and flow properties. Here it is assumed that if a certain flow velocity is specified, then the particles, the hovering velocity of which exceeds that of boundary-sized particles, settle down, whereas particles with hovering velocity below the boundary one ascend. This characteristic is unanimously considered among determining parameters of the process. The above analysis shows that it is far from being true. The value of particles hovering velocity by itself does not reflect the conditions of optimal mechanism of the process, since it does not take into account the design.

Channel design and boundary conditions play a certain role in the setting of correct regimes of optimal separation. This can be confirmed by the following example. Special studies were performed on a coaxial apparatus comprising 11 identical stages ($z=11$). The place of material feed in these experiments was successively changed from the upper ($i^*=1$) to the lower ($i^*=11$) stage. First, the flow velocity was chosen so that to obtain the optimal separation condition, i.e. $k=50\%$, at each stage. It was clarified how it affects the total extraction of a narrow-sized class for the entire apparatus at different places of the initial material feed. Then an inverse problem was solved. A flow regime providing optimal separation for a fixed class in the entire apparatus volume, i.e. $F_f(x)=50\%$, was chosen, and fractional extraction for one stage was determined. The results of these studies are summarized in Table 12.4

Interesting conclusions follow.

First, hovering velocity ensures optimal separation by the desired boundary in one case only—at a medium (along the apparatus height) level of the material feed into the flow.

Second, for other feed levels, hovering conditions do not provide optimal separation. For upper levels of material feed, optimal flow velocities exceed hovering velocity, and for lower levels are below the latter.

This unambiguously points to the fact that optimal flow velocities are determined not only by its dynamic characteristics, but also, to a significant extent, by the apparatus design and boundary conditions, as it follows from Table 12.4.

Analysis of the results of the same experiments shows that maximal extraction of target fraction into the fine product corresponds to upper feed. With lowering of the feed level, the amount of target fraction in the coarse product grows. At the same time, fine product contamination is minimal at the medium feed level and increases with the rise of feed level.

Thus, all of this testifies that the “velocity hypothesis” generally accepted today in the theory of gravitational concentration does not stand up to criticism.

Annotation

In the present book, a number of invariants for two-phase flows are formulated, substantiated and experimentally tested from the standpoint of statistical mechanics. This solves complicate, theoretical and applied problems related to such processes. It should be acknowledged that the main achievement of this book is the determination of two-phase flows entropy. Entropy has allowed an in-depth understanding of the main regularities and provided comprehensive methods of computing mass-exchange processes in such flows. Entropy is a global invariant for all sections of modern physics, biology, chemistry and other sciences dealing with energy and matter transformations. Entropy has allowed a considerable progress in the understanding of mechanisms of two-phase flows.

The book is intended for students, including those working for Master's or Doctor's degrees in engineering sciences dealing with processes involving two-phase flows. It will be useful for engineers engaged in problems of power generation, chemical technology, heat engineering, raw minerals concentration, building, metallurgy, materials for electronics, etc., and even in perfumery and food industry.

Bibliography

1. Barsky E. *Critical regimes of two-phase flows with a polydisperse solid phase*. London–New York: Springer; 2010.
2. Barsky E, Barsky M. *Cascade separation of powders*. Cambridge International Science Publishing; 2006.
3. Prigogine I, Stenders I. *Time, chaos, quantum*. Moscow: URSS; 2005.
4. Feynman R. *Lectures in physics*. Moscow: Mir; 1967.
5. Morse P. *Thermal physics*. Moscow: Nauka; 1967.
6. Lorentz HA. *Lectures in thermodynamics*. Moscow: Gosgortekhizdat; 1946.
7. Prigogine I, Stenders I. *Order out of chaos*. Moscow: URSS; 2005.
8. Landau LD, Lifshitz EM. *Statistical physics*. Moscow: Nauka; 1964.
9. Reist PC. *Introduction to aerosol science*. New York: Macmillan Publishing Co; 1987.
10. Fuks NA. *Mechanics of continuous media*. Moscow: Akademizdat; 1955.
11. Barsky MD, Revnivtsev VI, Sokolkin YuV. *Gravitational classification of granular materials*. Moscow: Nedra; 1974.
12. Barsky MD. *Optimization of process of granular materials separation*. Moscow: Nedra; 1978.
13. Barsky MD. *Fractionating of powders*. Moscow: Nedra; 1980.
14. Prigogine I. *The end of certainty*. New York: the Free Press; 1997.
15. Rosen AM. *Theory of isotope separation in columns*. Moscow: Atomizdat; 1960.
16. Soo SL. *Fluid dynamics of multiphase systems*. Blaisdell Publishing Co; 1971.
17. Gorbis ZR. *Heat exchange and hydrodynamics of dispersed through flows*. Moscow: Energia; 1970.
18. Monin AS, Yaglom AM. *Statistical hydromechanics*, vol. 1. Moscow: Nauka; 1965. vol. 2, 1967.
19. Happel I, Brenner H. *Low Reynolds number hydrodynamics*. Prentice Hall; 1965.
20. Gibbs JW. *Elementary principles in statistical mechanics: developed with special reference to the rational foundation of thermodynamics*. Yale University Press; 1902.
21. Tolmon RC. *Principles of statistical mechanics*. Oxford University Press; 1938.
22. Kittel C. *Thermal physics*. New York: John Willey and Sons Inc; 1977.
23. Brilloun L. *Science and information theory*. New York: Academic Press, Inc.; 1977.
24. Chambodall PP. *Evolution et applications du concept d'entropie*. Paris: Dunov; 1963.
25. Muschelknautz E. *Theoretische und experimentalle Untersuchungen über die Druckverluste pneumatischer Förderleitungen unter besonderer unter besonderer Berucksichtigung des Einflusses von Gutreibung und Gutgewicht*. Dusseldorf: Forschung-geschelt; 1959. t. 25.
26. Gandmacher FR. *Applications of the theory of the matrices*. New York: John Wiley and Sons; 1959.
27. Kemeny J, Shell J. *Finite Markov chains*. New York: Princeton; 1967.
28. Kemeny J, Shell J, Knapp A. *Markov chains*. Springer-Verlag; 1976.
29. Kondepudi D, Prigogine I. *Modern thermodynamics*. New York: John Wiley and Sons; 2000.
30. Seader JD, Henley EJ. *Separation process principles*. New York: John Wiley and Sons; 1998.
31. Khony FM. *Predicting the performance of multistage separation process*. Boca Raton, Florida: CRC Press; 2000.
32. Einstein WG. *General course of chemical technology processes and apparatuses*, vol. 1, vol. 2, Moscow, Lotos; 2002.
33. Abramovich GN. *Applied gas dynamics*. Moscow: Nauka; 1976.
34. Gelperin NI. *Principal processes and apparatuses of chemical technology*. Moscow: Khimia; 1981.
35. Levich VG. *Physicochemical hydrodynamics*. Moscow: Fizmatgiz; 1959.
36. Zakgeim AYu. *Introduction to modeling chemico-technological processes*. Moscow: Khimia; 1982.
37. Billing CE. *Controlling airborne particles*. Washington: National Academy of Sciences; 1980.

38. Chapman S, Cowling TG. *The mathematical theory of non-uniform gases*. Cambridge University Press; 1961.
39. Einshtein A. *The theory of Brownian movement*. New York: Dover; 1956.
40. McDaniel EW. *Collision theory in ionized gases*. New York: Wiley; 1964.
41. Schliching H. *Boundary layer theory*. New York: McGraw-Hill; 1968.
42. Sedov LI. *Planar problems of hydrodynamics and aerodynamics*. Moscow: Nauka; 1966.
43. Govorov AV. Cascade and combined processes of bulk materials fractionating. [Thesis]. Sverdlovsk: Urals Polytechnic Institute; 1986.
44. Shishkin SF. Intensification of the process of gravitational pneumatic classification [Thesis]. Sverdlovsk: Urals Polytechnic Institute; 1983.
45. Fermi E. *Notes on thermodynamics and statistics*. Chicago University Press; 1966.
46. Barsky M, Barsky E. Criterion for efficacy of separation of a pourable material into N components. In: *Proceedings of the 29th international symposium on computer applications in the mineral industries*; 2001.
47. Barsky M, Barsky E. Algorithms for optimization of classification of a pourable material into N components. In: *Proceeding of the 29th international symposium on computer applications in the mineral industries*; 2001.
48. Barsky E. Theoretical basis for separation of pourable materials in vertical flows. *Sci Isr–Technol Advantages* 2003;(1).
49. Lorentz HA. *Les Theories statistiques en Thermodynamique*; 1916. Leipzig-Berlin.
50. Barsky E, Barsky M. Optimal air flow velocities in gravity separation processes. *Miner Process J–Enrich Ores, St. Petersburg* 2002;(2).
51. Fortier A. *Mechanics of suspensions*. Moscow: Mir; 1971.
52. Kay JM. *An introduction to fluid mechanics and heat transfer*. Cambridge University Press; 1957.
53. Barsky E, Barsky M, Govorov A. Lifting power and structure of two-phase flows in gravity separation regimes. *Sci Isr–Technol Advantages* 2002;**4**.
54. Volkenstein MV. *Configurational statistics of polymers*. Academy of Sciences of the USSR; 1959.
55. Gibbs JW. *The scientific papers of J. Willard Gibbs*, vol. 1. Dover; 1961. Thermodynamics. vol. 4.
56. Barsky E, Barsky M. Master curve of separation processes. *Physical Sep Sci Eng* 2004;(2).
57. Barsky E. Algorithms for efficacy of separation of pourable material. *ECMI Newsl Math Ind* 2004;(36).
58. Barsky E. Efficacy of separation of pourable material. *ECMI Newsl Math Ind* 2004;(36).
59. Barsky E. Conditions providing optimum separation. *Phys Sep Sci Eng* 2004;(3).
60. Barsky E. Absorbing Markov chain in gravitational cascade separation of pourable materials at different stages of a classifier. *ECMI Newsl Math Ind* 2005;(37).
61. Barsky E. Medium motions regimes and universal curves of gravitational separation. *Int J Miner Process* 2007;(3).
62. Barsky E, Barsky M. Correlation between the velocities of particle settling and the optimal stream velocity in gravity separation. 22nd international mineral processing congress. Cape Town; 2003.
63. Barsky E. Chaos and order in turbulent two-phase currents. In: *Proceeding of first international conference on discrete chaotic dynamics in nature and society*; 1998.
64. Sheinberg A. *Preparation of coarse dust of brown coal*. Moscow: Metallurgizdat; 1970.
65. Heisenberg W. *The physical principles of the quantum theory*. Chicago: University Press; 1930.
66. Mandelbrot B. *The fractal geometry of nature*. NY: Freeman; 1982.
67. De Leener M. *Classical kinetics of fluids*. NY: Wiley; 1977.
68. Prigogine I. *From being to becoming*. San Francisco: Freeman and Co; 1980.

Index

Note: Page numbers followed by f indicate figures; t, tables; b, boxes.

A

Affine dependence, 78, 78f
Affinization, 167
Apparatus hermetization, 236
Archimedes criterion, 44, 163
Archimedes volume, 117
Avogadro number, 4

B

Big Bang theory, 10
Binary mixture, 83
  heterogeneity, 88
Binomial distribution, 103
Boltzmann's theory, 9–10, 101, 123
Boundary class entropy, 157
Boundary layer, 38
  turbulization, 43
Boundary-size grain, 189–190

C

Canonic form, 193–194
Cascade separation
  absorbing Markov chain in, 193–195
  adequacy method, experimental check of, 232–235
  discrete model, 180–187
  initial material apparatus, cyclic feed of, 191–193
  main principles of, 175–180
  mathematical model of, 195–200
    analysis, 187–191
  multirow classifier operation, optimization of, 239–242
  multirow combined, computation method of, 235–238, 235f
  multi-stage separation, principal parameters of, 212–221
  perfection estimating, invariants for, 223–230
  vertical method, theoretical estimation of, 230–232
  vertical two-phase flows, critical regimes discrete model of, 200–211
Cascade separator, 61f
Chaotizing factor, 129
  value of, 157
Clausius principle, 7–8
Collision equations, 9
Conical apparatuses, 231
Continuous flows
  aspects of, 18–22
  flow parameters analysis, using similarity criteria, 28–36
  invariants for, 22–27
  model technological processes, substantiation of, 17–18
Continuous medium velocities, 168
Continuous regime, 204–205
Continuous separation process, 146
Criterial equations, 27, 27f

D

d'Alembert's paradox, 37–38
Dalton's law, 3, 155–156
Density distribution analysis, 82
Descartes theorem, 188
Discrete model, 180–187
Dispersed component, 98
Distribution coefficient, 168
Distribution function, 80
Diverse cascade classifiers, 223–224
Drag crisis, 43
Dynamic entropy component, 133, 140
Dynamic equilibrium, 29f

E

Efficiency, 79
Elementary layer, 38
Elevated velocity region, 31
Empirical invariants
  fractional extraction, 59–64, 61f
    affinization of, 64–68
  generalizing, 74–78
  unified separation curves, 68–74
Entropic stability, 160
Entropy
  connection, 117
  ensemble of states and probability, 7–10
  of mixture composition, 84
  problematic aspects of, 10–16
  substantiation, 1–7
Ergodic principle, 102
Experimental curves, 60
External interaction model, 98–101

F

Falling layer, 59, 119
Falling velocity. *See* Final settling velocity
Final settling velocities, 40, 43–45
Flow continuity equation, 20

Flow element, dynamic equilibrium of, 29f
Flow velocity, 161
  ensuring optimal separation, 243–246, 246t
Formulated entropy criterion, 149
Fractional extraction, 167
Fractional separation, 221f
Friction coefficients, 100
Friction force, 100
Froude criterion, 73

G
Gas molecules, 7
Gauss distribution, 111
Gay-Lussac's law, 14
Granulometric composition, 74, 157
  invariants for, 79–83
    polyfractional mixture, 80–81, 81f
Gravitational fields, 130

H
Hamiltonian, 121–122
Heat death theory, 2
Hollow (equilibrium) apparatuses, 231
Hydraulically smooth pipes, 34
Hydrodynamic pattern, 39
Hyperbolic tangent, 40

I
Impact interactions, 94
Initial material apparatus, cyclic feed of, 191–193
Intermediate regime, 166
Irreversible processes, 156
Isolated separation act (ISA), 201–202

K
Kinematic coagulation, 53
Kinetic energy molecules, 7
Kinetic theory, 93

L
Lagrange factor, 142
Laminar overflow regime, 45
Le Chatelier principle, 6
Lifting factor, 112
Lower branch extension, 225

M
Magnus effect, 46
Markov chain, 193–195
Mass distribution process, stability and, 155–160
Mass precipitation, 50f
Mass suspension, 52–56
Mass transfer
  generalising invariants for parameters, 130–133
  maximal entropy condition, 140–142
  multidimensional model, 121–126
  polyfractional solid phase, 133–140
  separation regimes, 126–130
Material distribution, 59, 60f, 179
Mathematical model analysis, 187–191, 195–200
Maximal entropy, 140–142
Maximal entropy value, 149
Mean momentum, 99
Measuring system, 56
Medium flow
  rate, 130
  velocity, 47
Minimal transfer velocity, 56
Mixed turbulent regime, 34
Mixture composition
  granulometric composition, invariants for, 79–83
  uncertainty of, 83–90
Modern notions
  carrying capacity of, 56–58
  features of, 49–52
  final settling velocities, 43–45
  mass suspension of particles in, 52–56
  moving medium, particle interaction with, 46–49
  unit particles settling, peculiarities of, 37–43
Monotonic velocity, 38
Multicomponent mixtures, 83
Multidimensional model, 121–126
Multidimensional system, 121
Multiproduct separation, 150, 150f, 153t
Multirow separator, 235f, 236
Multi-stage separation, principal parameters of, 212–221

N
Narrow-class distribution, 176–177, 177f–178f
Narrow-class particles distribution, 176, 176f
Navier–Stokes equations, 21
Newton–Rittinger law, 162
Newton's law, 44
Nonequilibrium processes, 157–158
Nonuniform cascade scheme, 196f

O
Optimization condition, 150
Optimization parameter, 149
Optimum distribution, 215

P
Parameter φ, 8

Particle collisions model, 94–98
Particle distribution, statistical equilibrium for, 148–155
Particle interaction, 45–49, 45f
Particles collisions, 62
Particles densities, bulk material distribution in, 81–82, 82f
Particle velocity distribution, 38f
Periodic regime, 205–206
Phase trajectory, 123
Physical analogy, substantiation of, 91–94
Pipe quality, 33
Pneumotransport regime, 119
Poly-fractional bulk material, 200–201
Polyfractional material mass distribution, 71
Polyfractional mixture, 139–140
Polyfractional solid phase, 119–120
Polyfractional system, 133
Potential energy, 132
Potential extraction, 112–113
Probabilistic analysis, 93, 105–106
Probabilistic notions, 121–122
Probability, 80
  density, 121–123
Proportional model reflecting mass exchange, 175, 176f
Pulsational velocity component distribution, 35
Pulsations frequency, 35

Q

Quadratic resistance law, 133
Quality criterion, 79
Quartzite classification, 233–234, 234f

R

Radial component, 98
Random process, 84
Reliable experimental data, 214
Reynolds number, 41, 41f
Reynolds' similarity law, 35

S

Self-similarity, 116
  region, 27, 27f, 49
Self-similar quadratic turbulent regime, 34
Semi-logarithmic dependence, 224, 224f
Semi-logarithmic interpretation, 64
Separating capacity, 229–230, 230f
Separation completeness parameter, 226, 227f
Separation curves, 60, 61f, 71f, 220–221, 221f
  particle extraction, 167–173
  solid-phase particle motion regimes, 161–167
Separation efficiency, 146
Separation factor, 216
Separation parameter value, 112
Separation regimes, 126–130
Separator designs, 64, 67f
Separator efficiency rank, 242–243, 243t
Settling velocity, 44
Single-phase pseudo-homogeneous medium, 54
Solid-particle velocity, 139
Solid-phase composition, 143–144, 144f
Solid-phase separation quality, 143–148
Solid polydisperse particles, 52
Specific properties
  particle distribution, statistical equilibrium for, 148–155
  solid-phase separation quality, 143–148
  stability and mass distribution process, 155–160
Stability, 155–160
Standard deviation, 103
Stationary process, 193
Statistical analysis, principal tools of, 93
Statistical identity, 132
Statistical parameters
  chaotising factor for, 117–120
  generalising invariant for, 113–117
  generalising parameter for, 106–113
  solid particles continuum, statistical system for, 105–106
Statistical system, 107, 107f
  external interaction model, 98–101
  particle collisions model, 94–98
  physical analogy, substantiation of, 91–94
  statistical model substantiation, 101–104
  substantiation, 101–104
Steady particle velocity, 47
Steady separation regime, 181
Stirling formula, 11, 85, 118
Stroboscopic photographs
  of particle motion, 63
Suspending coarse heavy particles, 53, 53f

T

Tangential forces, 20–21
Temperature ratio, 13
Thermodynamic entropy, 13
  fluctuation, 160
Thermodynamic equilibrium, 6
Thermodynamic system, 84
Three-component mixture, 88, 89f
  entropy surface for, 90, 90f
Total shock pulse, 97
Total solid-phase composition, 140
Turbulence intensity, 35
Turbulence scale, 35
Turbulent flows, 34
Turbulent motion, 45
Turbulent vortices generation, 31f

U

Unified curve, 70
Unit flow volume, 96

V

Velocity
  distribution, 28, 28f, 30
  hypothesis, 92–93
  transformation, 94–95, 95f
Vertical two-phase flows, critical regimes discrete model of, 200–211
Viscosity forces, 28

Z

Zigzag-type apparatus, 233, 233f

Kolibri 360 GmbH
Ferdinand-Jühlke-Straße 7
99095 Erfurt, Deutschland
produktsicherheit@kolibri360.de